Daniel Zivkovic

Neutral non-equilibrium population genetics

Daniel Zivkovic

Neutral non-equilibrium population genetics

Südwestdeutscher Verlag für Hochschulschriften

Impressum/Imprint (nur für Deutschland/ only for Germany)
Bibliografische Information der Deutschen Nationalbibliothek: Die Deutsche Nationalbibliothek verzeichnet diese Publikation in der Deutschen Nationalbibliografie; detaillierte bibliografische Daten sind im Internet über http://dnb.d-nb.de abrufbar.

Alle in diesem Buch genannten Marken und Produktnamen unterliegen warenzeichen-, marken- oder patentrechtlichem Schutz bzw. sind Warenzeichen oder eingetragene Warenzeichen der jeweiligen Inhaber. Die Wiedergabe von Marken, Produktnamen, Gebrauchsnamen, Handelsnamen, Warenbezeichnungen u.s.w. in diesem Werk berechtigt auch ohne besondere Kennzeichnung nicht zu der Annahme, dass solche Namen im Sinne der Warenzeichen- und Markenschutzgesetzgebung als frei zu betrachten wären und daher von jedermann benutzt werden dürften.

Verlag: Südwestdeutscher Verlag für Hochschulschriften Aktiengesellschaft & Co. KG
Dudweiler Landstr. 99, 66123 Saarbrücken, Deutschland
Telefon +49 681 37 20 271-1, Telefax +49 681 37 20 271-0
Email: info@svh-verlag.de
Zugl.: Köln, Universität zu Köln, Diss., 2008

Herstellung in Deutschland:
Schaltungsdienst Lange o.H.G., Berlin
Books on Demand GmbH, Norderstedt
Reha GmbH, Saarbrücken
Amazon Distribution GmbH, Leipzig
ISBN: 978-3-8381-1223-7

Imprint (only for USA, GB)
Bibliographic information published by the Deutsche Nationalbibliothek: The Deutsche Nationalbibliothek lists this publication in the Deutsche Nationalbibliografie; detailed bibliographic data are available in the Internet at http://dnb.d-nb.de.

Any brand names and product names mentioned in this book are subject to trademark, brand or patent protection and are trademarks or registered trademarks of their respective holders. The use of brand names, product names, common names, trade names, product descriptions etc. even without a particular marking in this works is in no way to be construed to mean that such names may be regarded as unrestricted in respect of trademark and brand protection legislation and could thus be used by anyone.

Publisher: Südwestdeutscher Verlag für Hochschulschriften Aktiengesellschaft & Co. KG
Dudweiler Landstr. 99, 66123 Saarbrücken, Germany
Phone +49 681 37 20 271-1, Fax +49 681 37 20 271-0
Email: info@svh-verlag.de

Printed in the U.S.A.
Printed in the U.K. by (see last page)
ISBN: 978-3-8381-1223-7

Copyright © 2010 by the author and Südwestdeutscher Verlag für Hochschulschriften Aktiengesellschaft & Co. KG and licensors
All rights reserved. Saarbrücken 2010

ZUSAMMENFASSUNG

Die Identifizierung genomischer Regionen, die von positiver darwinscher Selektion geprägt wurden, gilt als eines der zentralen Interessen der molekularen Evolutionsbiologie. Wenn eine Population ein neues Territorium besiedelt oder einer drastischen Umweltveränderung ausgesetzt ist, bedingt dies naturgemäß eine Schwankung in der Populationsgrösse. Dennoch beruhen die meisten statistischen Methoden, die adaptive Vorgänge in einer Population erfassen sollen, auf der Annahme einer konstanten Populationsgrösse. Erst in den letzten Jahren wurde Populationsgenetikern bewusst, dass Modelle variabler Populationsgrösse für eine angemessene Interpretation von DNA-Polymorphismen unerlässlich sind.

Im Rahmen der Koaleszenztheorie werden theoretische Resultate für die zweiten Momente bestimmter Baummaße unter variabler Populationsgröße hergeleitet. Darüber hinaus werden Formeln für die zweiten Momente diverser DNA-Polymorphismus-Maße für allgemeine binäre Bäume entwickelt. Diese Resultate stellen das Herzstück der Dissertation dar. Mit Hilfe dieser Ergebnisse erhält man tiefere Einsichten in die Auswirkung variabler Populationsgrösse und grenzt das Problem der Unterscheidung adaptiver und demographischer Faktoren besser ein.

Im Folgenden werden verschiedene weit verbreitete statistische Testmethoden, die ursprünglich unter der Annahme einer konstanten Populationsgrösse konstruiert wurden, verallgemeinert, sodass, wenn die demographische Entwicklung einer Population ausreichend bekannt ist, einzelne Loci gegen die Nullhypothese neutraler Evolution unter variabler Populationsgrösse getestet werden können. Dies wird anhand zweier X-chromosomaler Datensätze einer afrikanischen und einer europäischen Stichprobe von *Drosophila melanogaster* demonstriert.

Während das im Vorfeld vorgeschlagene Expansionsmodell für die afrikanische Stichprobe erfolgreich in die verallgemeinerten Teststatistiken integriert werden kann, bleibt eine adäquate Analyse für das Bottleneckmodell der europäischen Stichprobe unerfüllt. Abschließend werden charakteristische Merkmale demographischer Modelle vorgestellt, anhand welcher die Durchführbarkeit einer aussagekräftigen statistischen Analyse nachvollzogen werden kann.

Abstract

The identification of genomic regions that have been exposed to positive Darwinian selection is of major interest in evolutionary biology. Although adaptive processes are generally associated with populations that experience an environmental change or colonize a new habitat, statistical tests were commonly constructed on the assumption of constant population size. However, only in recent years did the practical need to account for models of variable population size become apparent in the attempts of population geneticists to properly interpret the rising amount of DNA polymorphism data.

Within the framework of coalescent theory, theoretical results regarding the second-order moments of certain tree size measures are derived under variable population size. Thereafter, formulas for the second-order moments of diverse DNA polymorphism measures are developed for general binary trees. These results constitute the centerpiece of this thesis. Their relevance lies in the possibility to obtain deeper insights into demographic factors and to better delimit the problem of distinguishing adaptive from demographic forces.

Several popular and widely applied statistical tests that rest on the assumption of constant population size are generalized, so that, conditional on knowledge of a demographic scenario, single loci can be tested for traces of selection against the null hypothesis of neutral evolution under variable population size. This is demonstrated for two datasets of X-linked loci from an African and an European sample of *Drosophila melanogaster*.

While a previously suggested expansion model for the African sample can be successfully implemented into the generalized test statistics, an adequate analysis for the bottleneck model of the European sample cannot be accomplished. Consequently, we extract characteristic features of demographic models in order to distinguish the ones which are accessible to a meaningful statistical analysis from those which are not.

CONTENTS

1. INTRODUCTION		1
1.1	Theoretical population genetics	1
1.2	Modeling neutral and adaptive mutations	4
1.3	Testing the neutral mutation hypothesis	6
1.4	Organization of the thesis	8
2. GENERAL COALESCENT TREES		11
2.1	Variable population size	12
	2.1.1 Mean waiting times	*13*
	2.1.2 Mean squared waiting times	*20*
	2.1.3 Mean product of two distinct waiting times	*25*
	2.1.4 Mean and variance of two tree size measures	*31*
2.2	Measures of DNA polymorphism	32
	2.2.1 The number of segregating sites	*32*
	2.2.2 Mutations of a certain size	*33*
	2.2.3 The average number of pairwise differences	*40*
3. TESTING NEUTRALITY UNDER VARIABLE POPULATION SIZE		41
3.1	The estimated demographic history of *Drosophila melanogaster*	41
3.2	Generalization of classical test statistics	45
	3.2.1 Tajima's D	*45*
	3.2.2 Fu and Li's D	*50*
	3.2.3 Fay and Wu's H	*54*

Contents

    3.2.4   The singleton-exclusive version of Tajima's D   57

4. DISCUSSION   **67**

REFERENCES   **77**

Chapter 1

INTRODUCTION

1.1 THEORETICAL POPULATION GENETICS

Population geneticists are devoted to the analysis of the evolutionary forces that shape the patterns of genetic variation within species. These forces include mutation, recombination, natural selection, demographic processes and their interactions. Theoretical population genetics is the mathematical study of these evolutionary patterns and forces. It is roughly divided into the development of conceptual models and theories (mostly based on the classical diffusion theory) and the advancements of statistical methods for data analysis (mostly based on coalescent theory).

Using the mathematical discipline of diffusion theory, classical population genetics is able to predict the impact of the evolutionary forces on allele or genotype frequencies at one or more loci of an entire population by looking forward in time. Once results for the whole population have been obtained, one can make predictions for a sample taken from the population, which is essential for the application of a chosen mathematical model to biological data. Initiated by the pioneering work of FISHER (1930b), WRIGHT (1931) and HALDANE (1932), diffusion models were state-of-art for decades, including important contributions by MALÉCOT (1948), FELLER (1951) and KIMURA (1964). Sewall Wright introduced the concept of genetic drift—the random alteration in allele frequency of a population over time. KIMURA (1968, 1983) established the neutral theory of molecular evolution. He argued that the vast majority of mutations are selectively neutral by having negligible effects on the reproductive ability of their carriers and that genetic drift is the primary evolutionary force—a hypothesis strongly opposed by GILLESPIE (1984, 2000). Despite the development of some major principles of the neutral theory (KIMURA 1983), the main advantage of the classical diffusion approach, in contrast to the coalescent-based approach described below, may be seen in the modeling of natural selection, with particular regard to positive selection. MAYNARD SMITH and HAIGH (1974) introduced the

1

Chapter 1. Introduction

model of genetic hitchhiking, in which a selectively neutral allele or mutation may quickly spread through an entire population if it is linked to a beneficial mutation. Consequently, the increase in frequency of the selectively neutral mutation leads to a drastic reduction in its variability—a procedure commonly known as a 'selective sweep'. Reformulating an approach of OHTA and KIMURA (1975), STEPHAN et al. (1992) obtained analytical results of the effect of a positive substitution on expected heterozygosity. However, addressing the same problem via coalescent theory is more cumbersome (KAPLAN et al. 1989).

Based on the seminal papers by KINGMAN (1982a, b), coalescent theory has become the most popular branch of theoretical population genetics over the past 25 years. In contrast to the classical diffusion approach, a sample (rather than the entire population) of genes, or DNA sequences, is traced backwards in time up to the single ancestor of the entire sample, which is referred to as the most recent common ancestor (MRCA). Figure 1.1 illustrates this ancestral process in form of a bifurcating genealogy. The intuitive treatment of population genetical questions has propelled coalescent theory to a broader readership (HEIN et al. 2005, WAKELEY 2008) due to the simpler acquisition of its basic principles, when compared to classical population genetics.

Another major advantage of this retrospective view of a sample in coalescent theory is given by the fact that computer simulations, which are often used to support mathematical approximations of specific models, are more time-efficient and easier to implement than classical, forward-in-time diffusion approaches.

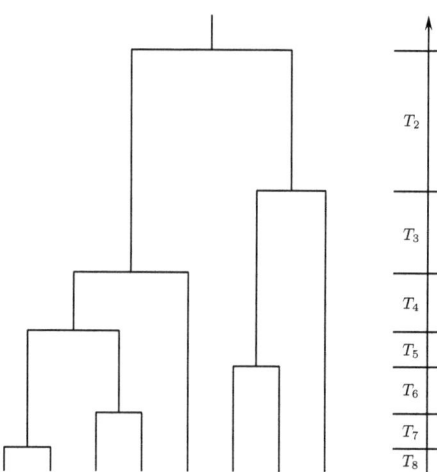

FIGURE 1.1   One possible coalescent tree of a sample of size eight. The waiting times until a pair of genes end up in their common ancestor are denoted by T_i.

1.1. Theoretical population genetics

ROSENBERG and NORDBORG (2002) reapplied two substantial arguments regarding the relationship of the sample to a large, unstructured population and the importance of recombination as an evolutionary force. First, as it has been shown by SAUNDERS et al. (1984), the probability that a sample of size n contains the MRCA of the entire population is simply $(n-1)/(n+1)$. For this reason, even the genealogy for a small sample is likely to contain the MRCA of the population and encourages a focus on the sample rather than on the population as a whole. Furthermore, this means that increasing the sample size has a minor effect. Adding more sequences in Figure 1.1 would mainly change the lower part of the genealogical tree, whereas the lengths of the deep branches near the MRCA would be barely affected (cf. HEIN et al. 2005). Second, the authors emphasized the importance of recombination for statistical inference. In the absence of recombination, there is only a single genealogical tree for an entire chromosome, such that a reliable prediction of its evolutionary history becomes arguable. In the presence of recombination, unlinked or loosely linked loci can be seen as independent replicates of the same ancestral process (cf. Figure 1.2). Therefore, variation sampled from a large number of preferably unlinked loci, which feature genealogies that are conditionally independent on the population's demographic history, provide the necessary polymorphism data for population genetical inferences. In particular, this shall lead to an improvement of the estimates of a population's demographic trend.

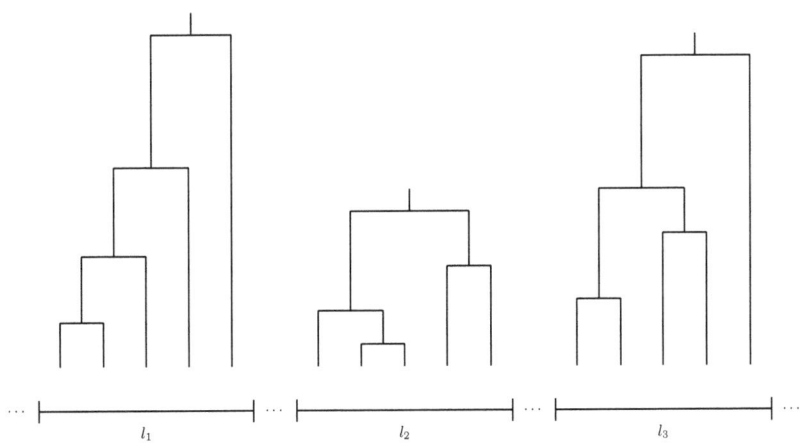

FIGURE 1.2   Possible coalescent trees for three different loci.

Chapter 1. Introduction

1.2 MODELING NEUTRAL AND ADAPTIVE MUTATIONS

Mutations are the indispensable source of variation to develop an evolutionary understanding of how demographic and selective forces have shaped present-day sampled DNA sequence variation data. Although these sequences reveal several types of genetic variation, such as copy number variations (REDON et al. 2006) or microsatellites (TAUTZ 1989), single nucleotide polymorphisms (SNPs) have become the prevalently analyzed form of polymorphism, especially from a theoretical standpoint. As such, only mutation models associated with SNPs are investigated throughout this thesis. The two most prominent assumptions regarding mutations are the infinitely-many-sites model (KIMURA 1969, WATTERSON 1975) and the infinitely-many-alleles model (KIMURA and CROW 1964). Under the infinitely-many-sites model, each new mutation arises at a site in an infinitely-long DNA sequence where there has never been a mutation before. The infinitely-many-alleles model, assuming that each mutation generates a new allele, differs from the infinitely-many-sites model in the sense that one ignores how many mutations distinguish alleles and considers only whether alleles are the same or different. The fundamental accomplishment under the infinitely-many-alleles model is the discovery of the Ewens sampling formula (EWENS 1972)—the probability distribution of a configuration of alleles in a sample of genes—under mutation-drift equilibrium. KARLIN and MCGREGOR (1972) were the first to prove the Ewens sampling formula. GRIFFITHS and LESSARD (2005) provided a proof by a direct combinatorial argument and extended the distribution to populations of varying size.

WATTERSON (1975) introduced the mutation parameter $\theta = 4N\mu$, where N is the size of a diploid population—at time of sampling in the coalescent theoretical context and, in particular, under variable population size—and μ is the mutation rate per sequence per generation. The number of mutations of rate μ or $\theta/2$, respectively, when time is measured in units of $2N$ generations, is typically Poisson-distributed in the sense that mutations are rare events. An important point in modeling selectively neutral mutations in the aforementioned manner is that these mutations have no impact on the structure of the genealogy of a sample. Consequently, the genealogical and mutational process can be separated; particularly when a population experiences changes in population size (see Chapter 2).

For the standard neutral model of constant population size, WATTERSON (1975) devised the mean and the variance of the number of segregating sites, S_n, which under the infinitely-many-sites model equals the number of mutations that occur on the genealogical tree. Later, TAVARÉ (1984) derived the probability distribution of S_n. Additionally, properties of other historically important test statistics, namely the average number of pairwise differences, Π_n, and the number of times a mutation can be observed at specific sites of the sampled sequences, ξ_i, have been derived within the coalescent framework. TAJIMA (1983) calculated the mean and the variance of Π_n and FU (1995) derived the means and the variances of ξ_i of selectively neutral alleles within a given sample under constant population size. Curiously, FISHER (1930a) already hinted at the result for the mean of ξ_i in the diffusion setting. The above results have been mostly obtained without taking recombination into account. The derivation of theoretical results

1.2. Modeling neutral and adaptive mutations

that incorporates recombination is more complex, even for the standard neutral model, as can be seen in KAPLAN and HUDSON (1985), who established an approximation of the variance of S_n, or WAKELEY (1997), who derived the variance of Π_n.

The study of temporal variation in population size has generated much interest since an early article by WRIGHT (1938). NEI et al. (1975) have analyzed the effect of arbitrary population size changes on the average heterozygosity of a neutral locus. MARUYAMA and FUERST (1984) developed a numerical method for the mean number and average age of alleles in a population undergoing an instantaneous expansion. WATTERSON (1984) derived analytical formulas, including the probability distribution and moments of the total number of alleles in a sample, for models of one or two sudden changes in population size. TAJIMA (1989a) derived the expected number of segregating sites and the expectation of the average number of pairwise differences for an instantaneous population growth model. SLATKIN and HUDSON (1991) examined a model of exponential growth before GRIFFITHS and TAVARÉ (1994) unified arbitrary changes in population size into a general framework. Moreover, GRIFFITHS and TAVARÉ (1998) developed the frequency spectrum of neutral alleles in a population of arbitrary varying population size.

Besides fluctuation in population size, the study of population structure has become another population genetical subfield of a long-standing interest ever since its initiation by Wright's island model (WRIGHT 1931), where a population is subdivided into discrete subpopulations with limited migration among them. Exemplarily, TAKAHATA and NEI (1985) derived the variance of Π_n in a two-population model without migration and WAKELEY (1996) obtained the concordant result with migration. WAKELEY (2008) provides a comprehensive overview on population structure, which is not further investigated in this thesis.

Since the different types of natural selection influence the reproductive success of the population's individuals, advantageous or deleterious mutations—in contrast to neutral mutations—have an effect on the genealogical tree. In addition to above work on positive directional selection, several advancements under the assumption of constant population size have been made. WIEHE and STEPHAN (1993) examined questions concerning the strength and the frequency of selective substitutions in the face of observable sequence data. BRAVERMAN et al. (1995) investigated the effect of genetic hitchhiking on the site-frequency spectrum by using a computer simulation based approach. FAY and WU (2000) pointed out that an excess of high-frequency derived variants is an unique characteristic of genetic hitchhiking. GILLESPIE (2000) introduced the pseudohitchhiking model—a simplification of the typically studied two-locus dynamics to a single locus—and argued that the recurrence of selected substitutions resembles the stochastic behavior of genetic drift. Whereas the above articles have been built on the initial suggestion by KAPLAN et al. (1989) that the number of individuals carrying the advantageous allele follows the logistic differential equation, there has been recent analytical and computational progress by making use of a so-called Yule approximation (SCHWEINSBERG and DURRETT 2005, ETHERIDGE et al. 2006). In contrast to the idea that adaptive substitutions are introduced by a new beneficial mutation in a single copy, HERMISSON and PENNINGS (2005) investigated the scenario where adaptive substitutions are derived from standing genetic variation.

Chapter 1. Introduction

OHTA (1973) suggested that, "very slight genetic deterioration might play an important role in molecular evolution". Selection against deleterious alleles, known as purifying selection, or background selection (CHARLESWORTH et al. 1993), is thought to be widespread across the genome (e.g., OHTA 1976). CHARLESWORTH et al. (1993) have demonstrated that background selection may have a similar effect on linked neutral polymorphism as directional selection; on the other hand purifying selection only induces slight changes to the frequency spectrum of linked neutral variants (e.g., PRZEWORSKI et al. 1999). Some research studies have addressed the joint effects of genetic hitchhiking and purifying selection on neutral variation under the standard model. These studies pointed out the reduction of the fixation probability of strongly selected alleles due to background selection (BARTON 1995) and demonstrated that formulas for background selection and hitchhiking can be combined to predict genetic variation at a linked neutral locus under constant population size (KIM and STEPHAN 2000).

Although it is reasonable to assume that adaptive mutations occur during environmental changes or when a population is colonizing a new habitat, theoretical studies have rarely considered the joint effect of demographic changes and selection until quite recently. WILLIAMSON et al. (2005) considered for the first time the combined effects of an instantaneous population size change and selection on the site-frequency spectrum. Furthermore, EVANS et al. (2007) studied the frequency spectrum of sites that are subject to selection and arbitrary population size changes within the framework of diffusion theory. However, the combined impact of positive directional selection and variations in population size on the frequency spectrum of a partially linked neutral locus remained unexplored so far.

1.3 TESTING THE NEUTRAL MUTATION HYPOTHESIS

The analysis and interpretation of patterns of DNA variation is the "great obsession" (GILLESPIE 2004) of molecular population geneticists. It is of particular interest to determine whether a locus is evolving neutrally or as a target of selection (HEY 1999). TAJIMA (1989b) was the first to construct a testable hypothesis of the standard neutral model, using only polymorphism data from within a population. Arguing that S_n and Π_n are unequally affected by the presence of selection, he combined the two different estimators of the mutation parameter, θ, based on S_n and Π_n, respectively, into a single test statistic, D, which is to this day commonly used throughout the scientific community. Subsequently introduced popular test statistics for deviation from neutral evolution (FU and LI 1993, FAY and WU 2000) posit a model of constant population size as well. KIM and STEPHAN (2002) proposed a composite-likelihood ratio (CLR) test to detect local signatures of genetic hitchhiking along a recombining chromosome for the standard neutral model. This was an improvement over previously proposed tests, since the null distribution is obtained by taking recombination into account. The main weakness of all these statistical tests lies in their inability to distinguish the effects of selection from demographic effects, such as changes in population size. For instance, JENSEN et al. (2005) demonstrated that the CLR test is not robust against certain

1.3. Testing the neutral mutation hypothesis

demographic scenarios. However, only in recent years has the practical need of non-equilibrium theories become evident in the attempts to disentangle selection from demography. In the model organism *Drosophila melanogaster*, several studies found evidence for the important role of demographic changes during the species' history (e.g., GLINKA et al. 2003, HADDRILL et al. 2005). Similarly, in humans (INTERNATIONAL HAPMAP CONSORTIUM 2005), the population size expansion that occurred after their migration out-of-Africa has required models that deviate from the standard equilibrium model of constant population size (e.g., WILLIAMSON et al. 2005, NIELSEN et al. 2005). Recently, several new methods and statistical tests were developed to analyze polymorphism data that presumably have been produced by a complex evolutionary history, during which selective and demographic forces acted simultaneously. NIELSEN et al. (2005) developed a composite likelihood method, which is an extension of the CLR test proposed by KIM and STEPHAN (2002), where the null hypothesis, rather than being fixed a priori, is derived from the pattern of background variation in the data. JENSEN et al. (2005) further proposed a goodness-of-fit test to accompany the CLR method (KIM and STEPHAN 2002).

The site-frequency spectrum has attracted a great deal of attention for the simultaneous inference of selection and demography. Based on the previously mentioned theoretical predictions for the site-frequency spectrum, WILLIAMSON et al. (2005) developed a maximum likelihood method and found evidence of population growth and purifying selection at non-synonymous sites in the human genome. LI and STEPHAN (2006) devised a maximum likelihood method, which allows inference of demographic changes and detection of recent positive selection in populations of varying size. Other popularized methods to detect traces of selection are centered around the structure and frequency spectrum of haplotypes (SABETI et al. 2002) or the analysis of linkage disequilibrium (LD), which is the non-random association of alleles at two or more loci. Besides demography, other neutral factors such as population subdivision and their consequent influence on LD may limit our abilities to detect signatures of selection in the genome. Nevertheless, several recently published articles showed the usefulness of LD for analyzing selective sweeps. KIM and NIELSEN (2004) have studied the patterns of LD caused by a selective sweep in a population of constant size. For this purpose, the authors investigated several established LD statistics by numerical simulations and proposed a new summary statistic, ω, which comprises the information of the square of the correlation coefficient in an allelic state (HILL and ROBERTSON 1968), r^2, between polymorphic sites. Furthermore, the authors have incorporated LD into the CLR test (KIM and STEPHAN 2002). The outcome that the addition of LD into the composite likelihood only slightly increases the power to detect selective sweeps was rather surprising. Recently, JENSEN et al. (2007) extended the numerical analysis of the closely related test statistic, ω_{max}, to non-equilibrium populations. In promising contrast, the authors demonstrated that for demographic parameters relevant to non-African populations of *D. melanogaster*, selected loci are distinguishable from neutral loci based on ω_{max}—with reasonable power—and further suggested that considering LD in conjunction with the site-frequency spectrum forms a valuable approach. STEPHAN et al. (2006) analyzed a deterministic three-locus model with one locus experiencing positive directional selection and two partially linked neutral loci and provided the surprising result that LD is completely eliminated after a selective

Chapter 1. Introduction

sweep, when the selected site is located between the neutral sites, suggesting that LD may indeed be used to pinpoint the target of selection. MCVEAN (2007) studied the effects of selective sweeps on patterns of LD by considering the relationship between LD and the structure of the underlying genealogy.

As an alternative to SNP-based methods, there are several approaches (e.g., SCHLÖTTERER 2002) to make use of microsatellites for the detection of selective sweeps. WIEHE et al. (2007) constructed a test statistic that considers variability patterns at multiple loci jointly. The reasoning here is that the traces of selective sweeps may be distinguished from those of recent population bottlenecks, since only the former have a local effect, while the latter should have a chromosome-wide effect. However, as a statistical test that is built on the assumption of constant population size, the usual range of demographic parameters that causes problems in distinguishing these two evolutionary forces, leads to a high rate of false-positives. Even more disillusioning, this test statistic may not be carried over into the non-equilibrium background. Already for the simplest stepwise mutation model (KIMURA and OHTA 1978) it is not clear how to theoretically separate the mutational from the genealogical process, since alleles are frequently identical by state without being identical by descent.

1.4 ORGANIZATION OF THE THESIS

GRIFFITHS and TAVARÉ (2003) wrote a highly influential article concerning "general coalescent trees", which may be seen as an extension of coalescent trees under variable population size. In this article, the authors summarize important theoretical results of the neutral theory, including the means of waiting times under variable population size, the frequency spectrum of a mutation and its age and the mean of the average number of nucleotide differences, Π_n, under general conditions. The first section of Chapter 2 builds on GRIFFITHS and TAVARÉ (1994). Here, we study the standard coalescent approximation to the Wright-Fisher model for a sample of n genes, without recombination and with population size varying in time. First, we revisit the calculation of the mean waiting times and present a promising recursive approach, which relies on conditional probabilities. In the following, this method is extended to resolve the second-order moments of waiting times, which must be discriminated into the mean of the squared waiting time, $E(T_k^2)$, and the mean of the product of two distinct waiting times, $E(T_{k'}T_k)$. While the variance of S_n immediately follows from the solutions of these statistical quantities, it is much more challenging to derive the variance of Π_n, $V(\Pi_n)$, and the covariance of S_n and Π_n, $\text{Cov}(S_n, \Pi_n)$, for general coalescent trees. To approach these two mathematical tasks, we follow the article of FU (1995), which particularly captures the analogous problems and provides their solutions, as corollaries of the second-order moments of the size of a mutation, under constant population size. All probabilities in Fu's work, which are related to the first- and second-order moments of waiting times, can be adapted to general coalescent trees. As for the remaining task, we generalize the formulas that join these genealogical properties with mutations, which are assumed to occur as independent Poisson processes along the edges of the tree, as in the articles of GRIFFITHS and TAVARÉ (1998, 2003).

1.4. Organization of the thesis

After the first and second-order moments of the size of a mutation are derived, it is routine to derive the formulas for $V(\Pi_n)$ and $\text{Cov}(S_n, \Pi_n)$ for general coalescent trees.
In Chapter 3, all revisited (GRIFFITHS and TAVARÉ 1998, 2003) and newly derived formulas regarding the first- and second-order moments of the size of a mutation and their corollaries are summarized into generalized versions of classical test statistics, which can be applied to test the neutral mutation hypothesis under variable population size to infer traces of selection. Since these test statistics assume an a priori knowledge of temporal changes in population size, it can only be applied if the necessary demographic parameters have been already estimated. LI and STEPHAN (2006) developed a maximum likelihood method to infer demographic changes and to detect recent selective sweeps in populations of varying size. The authors analyzed 262 and 272 X-linked loci from an African and a European population of *D. melanogaster*, respectively, and estimated their demographic histories based on the frequency spectra. First, we adapt the suggested demographic histories into the summary statistic D'—the generalization of Tajima's D (TAJIMA 1989b) to test the neutral mutation hypothesis under variable population size—and analyze to what extent this test statistic can be used to infer the traces of selection on single loci. INNAN and STEPHAN (2000) have addressed the same issue, using coalescent simulations in order to distinguish the effects of exponential growth and selection in *Arabidopsis thaliana*. In analogy to the standard neutral version of Tajima's D, the summary statistics of FU and LI (1993) are able to be tested against the neutral non-equilibrium model. Even more pronounced than Tajima's D, these test statistics measure an excess of singletons (i.e. site variants that occur once in a sample), which can either result from exponential growth (FU 1997), genetic hitchhiking (KIM and STEPHAN 2002), purifying selection (SMITH and EYRE-WALKER 2002) or possibly by sequencing errors (ACHAZ 2008) and should therefore be considered with caution. Exemplarily, we generalize one of these test statistics that relates the number of segregating sites, S_n, to the number of singletons, but, as in the demographic history of the African population of *D. melanogaster*, this generalization may incorporate the impact of exponential growth, but may not distinguish positive from purifying selection or manage certain technological issues. Thereafter, a generalized version of the H test (FAY and WU 2000) is constructed to particularly measure an excess of high-frequency derived variants. Due to the problematic interpretation of singletons, the chapter closes with a singleton-exclusive version of Tajima's D, which is established in analogy to a recently proposed standard neutral version (ACHAZ 2008).

It may be anticipated at this point, that the correction of the above mentioned test statistics for the African demography is well accomplishable, whereas the distributions of the standardized test statistics incorporating the European demography may be seen as too unsatisfactory for reliable inference. Interestingly, demographic parameter constellations, which were already critical for the use of test statistics that rely on the standard neutral model, remain cumbersome to some extent, when these are taken into account. The classification of delicate demographic scenarios marks the beginning of Chapter 4. We illustrate that the variance-to-mean ratio of total tree length (T_c), a simple measure for the distortion of the distribution of T_c and accordingly for S_n, can be applied to detect intractable population bottlenecks. Furthermore, we show that different models of population size change typically lead to different frequency spectra. This question is

Chapter 1. Introduction

particularly important to assess whether it is sufficient to estimate a single one out of numerously possible demographic parameter constellations. We conclude with an outline of possible future research questions.

GENERAL COALESCENT TREES

Throughout this chapter we investigate statistical measures, which reflect the ancestry of a sample of n genes, when recombination is absent. GRIFFITHS and TAVARÉ (1998, 2003) have established the theoretical framework for general coalescent trees, which are specified as follows. Let $T_n, \ldots, T_2$ be the time periods during which the coalescent tree has $n, \ldots, 2$ lineages, respectively. The general joint distribution of waiting times $(T_n, \ldots, T_2)$ satisfies the assumptions that:

(A1) $T_n, \ldots, T_2$ are continuous random variables.
(A2) The ancestral tree is binary, and such that when there are k ancestral lines each pair has probability $\binom{k}{2}^{-1}$ of being the next pair to coalesce.

To model the effects of mutation on general coalescent trees, it is assumed that:

(A3) Conditional on the edge lengths of the tree, mutations occur according to independent Poisson processes of rate $\theta/2$ along the edges of the tree.

While for the theoretical treatment of general coalescent trees generic time units are used, one time unit corresponds to $2N$ generations, where N is the current size of a diploid population, under variable population size. The compound parameter θ is given by $\theta = 4N\mu$, where μ is the mutation rate per sequence per generation. Therefore, $\theta/2$ is the mutation rate per sequence per $2N$ generations, providing the convenience of the same time scaling of the mutational and the genealogical process. Furthermore, mutations occur according to the infinitely-many-sites model, introduced earlier, under which the number of mutations in the genealogy of a sample of size n equals the number of segregating sites of these n sequences.

GRIFFITHS and TAVARÉ (1998, 2003) have derived numerous analytical results for

Chapter 2. General coalescent trees

general coalescent trees. Most notably, GRIFFITHS and TAVARÉ (1998, 2003) have derived the frequency spectrum, which is the probability distribution $q_{n,i}$ of the number of times, i, a single mutation arising between the present and the time of the most recent common ancestor, is represented in the sample of size n, as $\mu \to 0$. The probability $q_{n,i}$ is given by

$$q_{n,i} = \frac{(n-i-1)!(i-1)! \sum_{k=2}^{n-i+1} k(k-1)\binom{n-k}{i-1}E(T_k)}{(n-1)! \sum_{k=2}^{n} kE(T_k)}, \quad 0 < i < n. \tag{2.1}$$

Equation 2.1, which relates the probabilities $q_{n,i}$ to the expected lengths of coalescent waiting times T_k, may be used to illustrate the hierarchy within general coalescent trees. While the mean waiting times of T_k appear symbolically in Equation 2.1, the usage of Equation 2.2 provides an explicit result of the frequency spectrum under variable population size. For the standard neutral model, which represents the simplest special case of a general coalescent tree, replacing $E(T_k)$ by $\binom{k}{2}^{-1}$ results in

$$q_{n,i} = \frac{i^{-1}}{\sum_{k=1}^{n-1} k^{-1}}.$$

2.1 VARIABLE POPULATION SIZE

We measure time backwards in units of $2N$ generations, where N is the size of a diploid population at time of sampling. Define $\lambda_N(t) = N(t)/N$ as the ratio of the population sizes at time t in the past and the present. Furthermore, let $\lambda(t)$, which arises as the limit as $N \to \infty$ to ensure that the population size becomes large in each generation and is supposed to be strictly positive for all $t > 0$, be real and piecewise continuous. The population-size intensity function Λ is defined by

$$\Lambda(t) = \int_0^t \frac{1}{\lambda(u)} du.$$

We assume that $\Lambda(\infty) = \infty$, so that a sample of genes has a most recent common ancestor (MRCA) with probability 1.
The joint density (GRIFFITHS and TAVARÉ 1994) of $(T_n, \ldots, T_2)$ is

$$g(t_n, \ldots, t_2) = \prod_{j=2}^{n} \binom{j}{2} \frac{1}{\lambda(s_j)} \exp\{-\binom{j}{2}(\Lambda(s_j) - \Lambda(s_{j+1}))\},$$

for $0 \le t_n, \ldots, t_2 < \infty$, where $s_{n+1} = 0$, $s_n = t_n$, $s_j = t_j + \ldots + t_n$, $j = 2, \ldots, n-1$. Although the waiting times $T_n, \ldots, T_2$ are not independent, unlike in the case of constant population size, where $g(t_n, \ldots, t_2) = g(t_n) \ldots g(t_2)$, we will use the joint density and the law of iterated expectations to derive the first two moments of waiting times.

2.1. Variable population size

GRIFFITHS and TAVARÉ (1998) already obtained the means based on previous results from GRIFFITHS (1980) and TAVARÉ (1984). To emphasize the dependence of the first- and second-order moments of waiting times on the sample size n, waiting times and densities carry an index, n, unless the meaning is clear from the context.

2.1.1 Mean waiting times

First, we prove two lemmas.

Lemma 2.1.

$$\int_0^\infty t \frac{\binom{j}{2}}{\lambda(t)} \exp\{-\binom{j}{2} \int_0^t \frac{1}{\lambda(u)} du\} dt = \int_0^\infty \exp\{-\binom{j}{2} \int_0^t \frac{1}{\lambda(u)} du\} dt.$$

Proof.

$$\int_0^\infty t \frac{\binom{j}{2}}{\lambda(t)} e^{-\binom{j}{2} \int_0^t \frac{1}{\lambda(u)} du} dt = -\int_0^\infty t \left(\frac{d}{dt} e^{-\binom{j}{2} \int_0^t \frac{1}{\lambda(u)} du}\right) dt = \text{(integration by parts)}$$

$$\underbrace{-(\lim_{t \to \infty} t e^{-\binom{j}{2} \int_0^t \frac{1}{\lambda(u)} du} - 0)}_{=0} + \int_0^\infty e^{-\binom{j}{2} \int_0^t \frac{1}{\lambda(u)} du} dt.$$

$\square$

In the last line of the proof, the assumption of $\Lambda(\infty) = \infty$ is necessary to ensure that the limit for $t \to \infty$ is zero.

Lemma 2.2.

$$\int_0^\infty \int_0^\infty \frac{\binom{n+1}{2}}{\lambda(t_{n+1})} \exp\{-\binom{n+1}{2} \int_0^{t_{n+1}} \frac{1}{\lambda(u)} du\} \exp\{-\binom{j}{2} \int_{t_{n+1}}^{t_{n+1}+t} \frac{1}{\lambda(u)} du\} dt_{n+1} dt =$$

$$\frac{\binom{n+1}{2}}{\binom{n+1}{2} - \binom{j}{2}} \left(\int_0^\infty \exp\{-\binom{j}{2} \int_0^t \frac{1}{\lambda(u)} du\} dt - \int_0^\infty \exp\{-\binom{n+1}{2} \int_0^t \frac{1}{\lambda(u)} du\} dt \right).$$

Chapter 2. General coalescent trees

Proof.

$$\int_0^\infty \int_0^\infty \frac{\binom{n+1}{2}}{\lambda(t_{n+1})} e^{-\binom{n+1}{2}\int_0^{t_{n+1}} \frac{1}{\lambda(u)}du} e^{-\binom{j}{2}\int_{t_{n+1}}^{t_{n+1}+t} \frac{1}{\lambda(u)}du} dt_{n+1} dt = \text{(substitute } t_{n+1} + t \text{ by } s\text{)}$$

$$\int_0^\infty \int_0^s \frac{\binom{n+1}{2}}{\lambda(t_{n+1})} e^{-\binom{n+1}{2}\int_0^{t_{n+1}} \frac{1}{\lambda(u)}du} e^{-\binom{j}{2}\int_{t_{n+1}}^{s} \frac{1}{\lambda(u)}du} dt_{n+1} ds =$$

$$\frac{\binom{n+1}{2}}{\binom{n+1}{2}-\binom{j}{2}} \int_0^\infty \int_0^s \frac{\binom{n+1}{2}-\binom{j}{2}}{\lambda(t_{n+1})} e^{-(\binom{n+1}{2}-\binom{j}{2})\int_0^{t_{n+1}} \frac{1}{\lambda(u)}du} e^{-\binom{j}{2}\int_0^{s} \frac{1}{\lambda(u)}du} dt_{n+1} ds =$$

$$\frac{\binom{n+1}{2}}{\binom{n+1}{2}-\binom{j}{2}} \int_0^\infty e^{-\binom{j}{2}\int_0^s \frac{1}{\lambda(u)}du} \left(\int_0^s \frac{\binom{n+1}{2}-\binom{j}{2}}{\lambda(t_{n+1})} e^{-(\binom{n+1}{2}-\binom{j}{2})\int_0^{t_{n+1}} \frac{1}{\lambda(u)}du} dt_{n+1} \right) ds =$$

$$\frac{\binom{n+1}{2}}{\binom{n+1}{2}-\binom{j}{2}} \int_0^\infty e^{-\binom{j}{2}\int_0^s \frac{1}{\lambda(u)}du} \left(\int_0^s (-\frac{d}{dt_{n+1}}e^{-(\binom{n+1}{2}-\binom{j}{2})\int_0^{t_{n+1}} \frac{1}{\lambda(u)}du}) dt_{n+1} \right) ds =$$

$$\frac{\binom{n+1}{2}}{\binom{n+1}{2}-\binom{j}{2}} \int_0^\infty e^{-\binom{j}{2}\int_0^s \frac{1}{\lambda(u)}du} \left(1 - e^{-(\binom{n+1}{2}-\binom{j}{2})\int_0^s \frac{1}{\lambda(u)}du} \right) ds = \text{(rename } s \text{ by } t\text{)}$$

$$\frac{\binom{n+1}{2}}{\binom{n+1}{2}-\binom{j}{2}} \left(\int_0^\infty e^{-\binom{j}{2}\int_0^t \frac{1}{\lambda(u)}du} dt - \int_0^\infty e^{-\binom{n+1}{2}\int_0^t \frac{1}{\lambda(u)}du} dt \right).$$

$\square$

The following chart illustrates the setup of recursions and the way of inductive reasoning.

$$
\begin{array}{ccccccc}
E(T_2)_2 & & & & & & \\
& \searrow & & & & & \\
E(T_3)_3 & & E(T_2)_3 & & & & \\
& \searrow & & \searrow & & & \\
E(T_4)_4 & & E(T_3)_4 & & E(T_2)_4 & & \\
& \searrow & & \searrow & & \searrow & \\
\cdots & & \cdots & & \cdots & & \cdots \\
& & E(T_k)_n & & & & \\
& & & \searrow & & & \\
& & & & E(T_k)_{n+1} & &
\end{array}
$$

2.1. Variable population size

Now, we start to successively derive $E(T_n)_n$, $E(T_{n-1})_n$, ... and perform these recursions in particular for small sample sizes. From the joint density above one immediately obtains

$$E(T_n)_n = \int_0^\infty t_n g_n(t_n)dt_n = \int_0^\infty t\frac{\binom{n}{2}}{\lambda(t)}\exp\{-\binom{n}{2}\int_0^t \frac{1}{\lambda(u)}du\}dt$$
$$= \int_0^\infty \exp\{-\binom{n}{2}\int_0^t \frac{1}{\lambda(u)}du\}dt,$$

by making use of Lemma 2.1. This result forms the initial step of the induction proof and provides in particular the solutions for $E(T_2)_2$ and $E(T_3)_3$. To calculate $E(T_2)_3$ we use conditional expectations:

$$E(T_2)_3 = E(E(T_2|T_3)_3) = \int_0^\infty g_3(t_3)E(T_2|T_3 = t_3)_3 dt_3$$
$$= \int_0^\infty \frac{\binom{3}{2}}{\lambda(t_3)}\exp\{-\binom{3}{2}\int_0^{t_3}\frac{1}{\lambda(u)}du\}\left(\int_0^\infty \exp\{-\binom{2}{2}\int_{t_3}^{t_3+t}\frac{1}{\lambda(u)}du\}dt\right)dt_3.$$

Note that the result for $E(T_2)_2$ is used on the right-hand side of the above equation. Due to conditioning on $T_3 = t_3$, the integration bounds are shifted by t_3. Using Lemma 2.2 we can continue and write

$$E(T_2)_3 = \frac{\binom{3}{2}}{\binom{3}{2}-\binom{2}{2}}\left(\int_0^\infty \exp\{-\binom{2}{2}\int_0^t\frac{1}{\lambda(u)}du\}dt - \int_0^\infty \exp\{-\binom{3}{2}\int_0^t\frac{1}{\lambda(u)}du\}dt\right).$$

In the same way, we obtain

$$E(T_{n-1})_n = \sum_{j=n-1}^n (-1)^{j+n-1}\alpha_{n,j,n-1}\int_0^\infty \exp\{-\binom{j}{2}\int_0^t \frac{1}{\lambda(u)}du\}dt,$$

where

$$\alpha_{n,j,n-1} = \frac{\binom{n}{2}}{\binom{n}{2}-\binom{n-1}{2}}.$$

Since this result provides us with the solution for $E(T_3)_4$, we finally derive $E(T_2)_4$ by conditioning the result for $E(T_2)_3$ on the newly introduced waiting time T_4.

$$E(T_2)_4 = E(E((T_2)_3|T_4)_4) = \int_0^\infty g_4(t_4)E((T_2)_3|T_4 = t_4)_4 dt_4 = \cdots$$

Chapter 2. General coalescent trees

$$
\begin{aligned}
&= \frac{\binom{3}{2}}{\binom{3}{2}-\binom{2}{2}} \frac{\binom{4}{2}}{\binom{4}{2}-\binom{2}{2}} \int_0^\infty \exp\{-\binom{2}{2}\int_0^t \frac{1}{\lambda(u)}du\}dt \\
&\quad - \frac{\binom{3}{2}}{\binom{3}{2}-\binom{2}{2}} \frac{\binom{4}{2}}{\binom{4}{2}-\binom{3}{2}} \int_0^\infty \exp\{-\binom{3}{2}\int_0^t \frac{1}{\lambda(u)}du\}dt \\
&\quad + \frac{\binom{3}{2}}{\binom{3}{2}-\binom{2}{2}} \Big(\frac{\binom{4}{2}}{\binom{4}{2}-\binom{3}{2}} - \frac{\binom{4}{2}}{\binom{4}{2}-\binom{2}{2}} \Big) \int_0^\infty \exp\{-\binom{4}{2}\int_0^t \frac{1}{\lambda(u)}du\}dt \\
&= \sum_{j=2}^{4} (-1)^{j+2} \alpha_{4,j,2} \int_0^\infty \exp\{-\binom{j}{2}\int_0^t \frac{1}{\lambda(u)}du\}dt.
\end{aligned}
$$

In analogy,

$$ E(T_{n-2})_n = \sum_{j=n-2}^{n} (-1)^{j+n-2} \alpha_{n,j,n-2} \int_0^\infty \exp\{-\binom{j}{2}\int_0^t \frac{1}{\lambda(u)}du\}dt, $$

where

$$ \alpha_{n,j,n-2} = \frac{\binom{n-1}{2}}{\binom{n-1}{2}-\binom{n-2}{2}} \frac{\binom{n}{2}}{\binom{n}{2}-\binom{j}{2}}, \quad j = n-1, n-2, $$

and $\alpha_{n,n,n-2} = \alpha_{n,n-1,n-2} - \alpha_{n,n-2,n-2}$.

When performing the next recursion step to calculate $E(T_{n-3})_n$, one might already see that

$$ \alpha_{n,k,k} = \prod_{i=k+1}^{n} \frac{\binom{i}{2}}{\binom{i}{2}-\binom{k}{2}} = \frac{(2k-1)!n!(n-1)!}{k!(k-1)!(n-k)!(n+k-1)!}. $$

The solutions of $\alpha_{n,k,k}$, $\alpha_{n,k+1,k}$, ... eventually lead to

$$ E(T_k) = \sum_{j=k}^{n} (-1)^{j+k} \alpha_{n,j,k} \int_0^\infty t\, g_j(t)dt, \quad 2 \le k \le n, \tag{2.2} $$

where

$$ \alpha_{n,j,k} = \frac{(2j-1)n!(n-1)!(k+j-2)!}{(j-k)!k!(k-1)!(n-j)!(n+j-1)!}, $$

$$ g_j(t) = \frac{\binom{j}{2}}{\lambda(t)} \exp\{-\binom{j}{2}\int_0^t \frac{1}{\lambda(u)}du\}, $$

and e.g., $g_2(t_2)$ is the density of $(T_2)_2$. The proof of Equation 2.2 requires the following combinatorial identity.

2.1. Variable population size

Lemma 2.3.

$$\sum_{j=k}^{n}(-1)^{j+k}\alpha_{n,j,k} = 0.$$

Proof.

$$\begin{aligned}
\sum_{j=k}^{n}(-1)^{j+k}\alpha_{n,j,k} &= \sum_{j=k}^{n}(-1)^{j+k}(2j-1)\frac{n!(n-1)!(k+j-2)!}{k!(k-1)!(n-j)!(n+j-1)!(j-k)!} \\
&= \frac{\binom{2k-2}{k-2}}{(k-1)\binom{2n-1}{n-1}}\sum_{j=k}^{n}(-1)^{j+k}(2j-1)\binom{2n-1}{n-j}\binom{k+j-2}{j-k}
\end{aligned}$$

(upper negation of last term)

$$\begin{aligned}
&= \frac{\binom{2k-2}{k-2}}{(k-1)\binom{2n-1}{n-1}}\sum_{j=k}^{n}(-1)^{j+k}(2j-1)\binom{2n-1}{n-j}(-1)^{j-k}\binom{1-2k}{j-k} \\
&= \frac{\binom{2k-2}{k-2}\binom{2(n-k)}{n-k}}{(k-1)\binom{2n-1}{n-1}}\sum_{j=k}^{n}((2j-2k)+(2k-1))\frac{\binom{2n-1}{n-j}\binom{1-2k}{j-k}}{\binom{2(n-k)}{n-k}} \\
&= \frac{\binom{2k-2}{k-2}\binom{2(n-k)}{n-k}}{(k-1)\binom{2n-1}{n-1}} \times \\
&\qquad \left(2\sum_{j=k}^{n}(j-k)\frac{\binom{2n-1}{n-j}\binom{1-2k}{j-k}}{\binom{2(n-k)}{n-k}} + (2k-1)\sum_{j=k}^{n}\frac{\binom{2n-1}{n-j}\binom{1-2k}{j-k}}{\binom{2(n-k)}{n-k}}\right) \\
&= \frac{\binom{2k-2}{k-2}\binom{2(n-k)}{n-k}}{(k-1)\binom{2n-1}{n-1}} \times \\
&\qquad \left(2(n-k)\frac{1-2k}{2(n-k)}\underbrace{\sum_{j=k}^{n}\frac{\binom{2n-1}{n-j}\binom{-2k}{j-k-1}}{\binom{2(n-k)-1}{n-k-1}}}_{=1} + (2k-1)\underbrace{\sum_{j=k}^{n}\frac{\binom{2n-1}{n-j}\binom{1-2k}{j-k}}{\binom{2(n-k)}{n-k}}}_{=1}\right) \\
&= 0.
\end{aligned}$$

$\square$

In the penultimate line of the proof of Lemma 2.3, both sums are equal to one because of Vandermonde's convolution. Now we are ready to complete the proof of Equation 2.2.

Chapter 2. General coalescent trees

Proof of Equation 2.2.

To infer $E(T_k)_{n+1}$ from the induction assumption given in Equation 2.2, we write

$$\begin{aligned}
E(T_k)_{n+1} &= E(E(T_k|T_{n+1})_{n+1}) = \int_0^\infty g_{n+1}(t_{n+1})E(T_k|T_{n+1}=t_{n+1})_{n+1}dt_{n+1} \\
&= \int_0^\infty \frac{\binom{n+1}{2}}{\lambda(t_{n+1})} \exp\{-\binom{n+1}{2}\int_0^{t_{n+1}} \frac{1}{\lambda(u)}du\} E(T_k)_n^{\lambda_s} dt_{n+1},
\end{aligned}$$

where (cf. Figure 2.1)

$$\begin{aligned}
E(T_k)_n^{\lambda_s} &= \text{(induction assumption)} \\
&= \sum_{j=k}^n (-1)^{j+k} \alpha_{n,j,k} \int_0^\infty t \frac{\binom{j}{2}}{\lambda_s(u)} \exp\{-\binom{j}{2}\int_0^t \frac{1}{\lambda_s(u)}du\} dt \\
&= \text{(Lemma 1)} \sum_{j=k}^n (-1)^{j+k} \alpha_{n,j,k} \int_0^\infty \exp\{-\binom{j}{2}\int_0^t \frac{1}{\lambda_s(u)}du\} dt \\
&= (\lambda_s(u) = \lambda(u+t_{n+1})) \sum_{j=k}^n (-1)^{j+k} \alpha_{n,j,k} \int_0^\infty \exp\{-\binom{j}{2}\int_0^t \frac{1}{\lambda(u+t_{n+1})}du\} dt \\
&= \text{(substitute } v = u + t_{n+1}) \sum_{j=k}^n (-1)^{j+k} \alpha_{n,j,k} \int_0^\infty \exp\{-\binom{j}{2}\int_{t_{n+1}}^{t_{n+1}+t} \frac{1}{\lambda(v)}dv\} dt \\
&= \text{(rename } v \text{ by } u) \sum_{j=k}^n (-1)^{j+k} \alpha_{n,j,k} \int_0^\infty \exp\{-\binom{j}{2}\int_{t_{n+1}}^{t_{n+1}+t} \frac{1}{\lambda(u)}du\} dt.
\end{aligned}$$

Therefore,

$$\begin{aligned}
E(T_k)_{n+1} &= \int_0^\infty \frac{\binom{n+1}{2}}{\lambda(t_{n+1})} e^{-\binom{n+1}{2}\int_0^{t_{n+1}} \frac{1}{\lambda(u)}du} \left(\sum_{j=k}^n (-1)^{j+k} \alpha_{n,j,k} \int_0^\infty e^{-\binom{j}{2}\int_{t_{n+1}}^{t_{n+1}+t} \frac{1}{\lambda(u)}du} dt \right) dt_{n+1} \\
&= \sum_{j=k}^n (-1)^{j+k} \alpha_{n,j,k} \int_0^\infty \int_0^\infty \frac{\binom{n+1}{2}}{\lambda(t_{n+1})} e^{-\binom{n+1}{2}\int_0^{t_{n+1}} \frac{1}{\lambda(u)}du} e^{-\binom{j}{2}\int_{t_{n+1}}^{t_{n+1}+t} \frac{1}{\lambda(u)}du} dt_{n+1} dt.
\end{aligned}$$

2.1. Variable population size

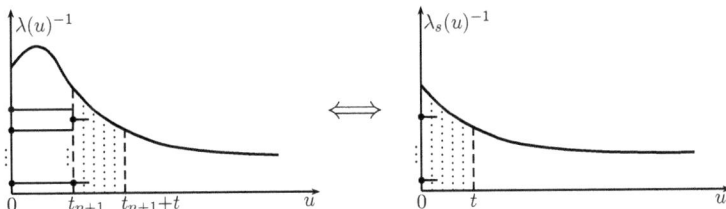

FIGURE 2.1 Since t_{n+1} is fixed within $E(T_k|T_{n+1} = t_{n+1})$, we can shift λ^{-1} by t_{n+1} to the left resulting in the function λ_s^{-1} with $\lambda_s(u)^{-1} = \lambda(u+t_{n+1})^{-1}$. Without loss of generality, we can use Equation 2.2 with respect to λ_s^{-1} (denoted as $E(T_k)_n^{\lambda_s}$) as our induction assumption as well.

By Lemma 2.2, we get

$$E(T_k)_{n+1} = \sum_{j=k}^{n}(-1)^{j+k}\alpha_{n,j,k}\underbrace{\frac{\binom{n+1}{2}}{\binom{n+1}{2}-\binom{j}{2}}}_{=\alpha_{n+1,j,k}}(\int_0^\infty e^{-\binom{j}{2}\int_0^t \frac{1}{\lambda(u)}du}dt - \int_0^\infty e^{-\binom{n+1}{2}\int_0^t \frac{1}{\lambda(u)}du}dt).$$

The usage of Lemma 2.3 results in

$$\begin{aligned}
E(T_k)_{n+1} &= \sum_{j=k}^{n}(-1)^{j+k}\alpha_{n+1,j,k}\int_0^\infty e^{-\binom{j}{2}\int_0^t \frac{1}{\lambda(u)}du}dt - \\
&\quad \Big(\underbrace{\sum_{j=k}^{n+1}(-1)^{j+k}\alpha_{n+1,j,k}}_{=0} - (-1)^{n+1+k}\alpha_{n+1,n+1,k}\Big)\int_0^\infty e^{-\binom{n+1}{2}\int_0^t \frac{1}{\lambda(u)}du}dt \\
&= \sum_{j=k}^{n+1}(-1)^{j+k}\alpha_{n+1,j,k}\int_0^\infty e^{-\binom{j}{2}\int_0^t \frac{1}{\lambda(u)}du}dt.
\end{aligned}$$

□

The expected waiting times for specific demographic scenarios can be derived from Equation 2.2. Population genetical models which consider a finite number of instantaneous population size changes can be explicitly solved, as seen in the following example of a population bottleneck (cf. Figure 2.2). Let $\lambda(t) = f$ for $\tau \leq t < \tau + \tau_f$ and $\lambda(t) = 1$ otherwise. Then,

$$E(T_k) = \sum_{j=k}^{n}(-1)^{j+k}\alpha_{n,j,k}\frac{1}{\binom{j}{2}}(1-(1-f)(\exp\{-\binom{j}{2}\tau\}-\exp\{-\binom{j}{2}(\tau+\frac{\tau_f}{f})\})),\ 2 \leq k \leq n.$$

Chapter 2. General coalescent trees

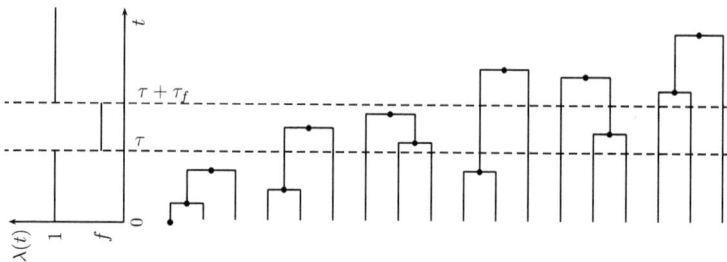

FIGURE 2.2   Possible coalescent trees for a 3-phase bottleneck model and sample size $n = 3$.

In the case of exponential growth, where $\lambda(t) = \exp\{-\beta t\}$ and $\beta > 0$, the mean waiting times $E(T_k)$ can be expressed in terms of the *exponential integral* $Ei(.)$ and are

$$E(T_k) = -\sum_{j=k}^{n}(-1)^{j+k}\alpha_{n,j,k}\frac{1}{\beta}\exp\{\frac{\binom{j}{2}}{\beta}\}\operatorname{Ei}(-\frac{\binom{j}{2}}{\beta}),\ 2 \leq k \leq n.$$

Obviously, the assumption of a large population size in each generation is violated at some time in the past. However, the upper equation yields the same result as if we would perform a logarithmic time-scale transformation (GRIFFITHS and TAVARÉ 1998) to the Kingman coalescent (KINGMAN 1982b). Models of instantaneous population size change followed by exponential growth can be expressed in terms of tabulated functions as well.

2.1.2 MEAN SQUARED WAITING TIMES

The same ideas of using conditional expectations and a proof by induction can be applied to the second-order moments of waiting times. To derive the mean squared waiting times $E(T_k^2)$, we first show the following Lemma.

Lemma 2.4.

$$\int_0^\infty\int_0^\infty\int_0^\infty t^2 \frac{\binom{n+1}{2}}{\lambda(t_{n+1})} \frac{\binom{j}{2}}{\lambda(t_{n+1}+t')} \frac{\binom{k}{2}}{\lambda(t_{n+1}+t'+t)} e^{-\binom{n+1}{2}\int_0^{t_{n+1}}\frac{1}{\lambda(u)}du} e^{-\binom{j}{2}\int_{t_{n+1}}^{t_{n+1}+t'}\frac{1}{\lambda(u)}du} e^{-\binom{k}{2}\int_{t_{n+1}+t'}^{t_{n+1}+t'+t}\frac{1}{\lambda(u)}du} dt_{n+1}dt'dt =$$

$$\frac{\binom{n+1}{2}}{\binom{n+1}{2}-\binom{j}{2}}\int_0^\infty\int_0^\infty t^2 \frac{\binom{j}{2}}{\lambda(t')} \frac{\binom{k}{2}}{\lambda(t'+t)} e^{-\binom{j}{2}\int_0^{t'}\frac{1}{\lambda(u)}du} e^{-\binom{k}{2}\int_{t'}^{t'+t}\frac{1}{\lambda(u)}du} dt'dt -$$

$$\frac{\binom{j}{2}}{\binom{n+1}{2}-\binom{j}{2}}\int_0^\infty\int_0^\infty t^2 \frac{\binom{n+1}{2}}{\lambda(t')} \frac{\binom{k}{2}}{\lambda(t'+t)} e^{-\binom{n+1}{2}\int_0^{t'}\frac{1}{\lambda(u)}du} e^{-\binom{k}{2}\int_{t'}^{t'+t}\frac{1}{\lambda(u)}du} dt'dt.$$

2.1. Variable population size

Proof.

$$\int_0^\infty \int_0^\infty \int_0^\infty t^2 \frac{\binom{n+1}{2}}{\lambda(t_{n+1})} \frac{\binom{j}{2}}{\lambda(t_{n+1}+t')} \frac{\binom{k}{2}}{\lambda(t_{n+1}+t'+t)} e^{-\binom{n+1}{2} \int_0^{t_{n+1}} \frac{1}{\lambda(u)} du} e^{-\binom{j}{2} \int_{t_{n+1}}^{t_{n+1}+t'} \frac{1}{\lambda(u)} du} e^{-\binom{k}{2} \int_{t_{n+1}+t'}^{t_{n+1}+t'+t} \frac{1}{\lambda(u)} du} dt_{n+1} dt' dt =$$

(substitute $t_{n+1} + t' + t$ by s_2 and $t_{n+1} + t'$ by s_1, respectively)

$$\int_0^\infty \int_0^{s_2} \int_0^{s_1} (s_2 - s_1)^2 \frac{\binom{n+1}{2}}{\lambda(t_{n+1})} \frac{\binom{j}{2}}{\lambda(s_1)} \frac{\binom{k}{2}}{\lambda(s_2)} e^{-\binom{n+1}{2} \int_0^{t_{n+1}} \frac{1}{\lambda(u)} du} e^{-\binom{j}{2} \int_{t_{n+1}}^{s_1} \frac{1}{\lambda(u)} du} e^{-\binom{k}{2} \int_{s_1}^{s_2} \frac{1}{\lambda(u)} du} dt_{n+1} ds_1 ds_2 =$$

$$\int_0^\infty \int_0^{s_2} (s_2 - s_1)^2 \frac{\binom{j}{2}}{\lambda(s_1)} \frac{\binom{k}{2}}{\lambda(s_2)} e^{-\binom{j}{2} \int_0^{s_1} \frac{1}{\lambda(u)} du - \binom{k}{2} \int_{s_1}^{s_2} \frac{1}{\lambda(u)} du} \frac{\binom{n+1}{2}}{\binom{n+1}{2} - \binom{j}{2}} \times$$

$$\underbrace{\left(\int_0^{s_1} \frac{\binom{n+1}{2} - \binom{j}{2}}{\lambda(t_{n+1})} e^{-\left(\binom{n+1}{2} - \binom{j}{2}\right) \int_0^{t_{n+1}} \frac{1}{\lambda(u)} du} dt_{n+1} \right)}_{= -\int_0^{s_1} \frac{d}{dt_{n+1}} e^{-\left(\binom{n+1}{2} - \binom{j}{2}\right) \int_0^{t_{n+1}} \frac{1}{\lambda(u)} du} dt_{n+1}} ds_1 ds_2 =$$

$$\int_0^\infty \int_0^{s_2} (s_2 - s_1)^2 \frac{\binom{j}{2}}{\lambda(s_1)} \frac{\binom{k}{2}}{\lambda(s_2)} e^{-\binom{j}{2} \int_0^{s_1} \frac{1}{\lambda(u)} du - \binom{k}{2} \int_{s_1}^{s_2} \frac{1}{\lambda(u)} du} \frac{\binom{n+1}{2}}{\binom{n+1}{2} - \binom{j}{2}} \left(1 - e^{-\left(\binom{n+1}{2} - \binom{j}{2}\right) \int_0^{s_1} \frac{1}{\lambda(u)} du}\right) ds_1 ds_2 =$$

$$\frac{\binom{n+1}{2}}{\binom{n+1}{2} - \binom{j}{2}} \int_0^\infty \int_0^{s_2} (s_2 - s_1)^2 \frac{\binom{j}{2}}{\lambda(s_1)} \frac{\binom{k}{2}}{\lambda(s_2)} e^{-\binom{j}{2} \int_0^{s_1} \frac{1}{\lambda(u)} du - \binom{k}{2} \int_{s_1}^{s_2} \frac{1}{\lambda(u)} du} ds_1 ds_2 -$$

$$\frac{\binom{j}{2}}{\binom{n+1}{2} - \binom{j}{2}} \int_0^\infty \int_0^{s_2} (s_2 - s_1)^2 \frac{\binom{n+1}{2}}{\lambda(s_1)} \frac{\binom{k}{2}}{\lambda(s_2)} e^{-\binom{n+1}{2} \int_0^{s_1} \frac{1}{\lambda(u)} du - \binom{k}{2} \int_{s_1}^{s_2} \frac{1}{\lambda(u)} du} ds_1 ds_2 =$$

(introduce $s_2 - s_1 = t$, rename s_1 by t', replace s_2 by $t' + t$)

$$\frac{\binom{n+1}{2}}{\binom{n+1}{2} - \binom{j}{2}} \int_0^\infty \int_0^\infty t^2 \frac{\binom{j}{2}}{\lambda(t')} \frac{\binom{k}{2}}{\lambda(t'+t)} e^{-\binom{j}{2} \int_0^{t'} \frac{1}{\lambda(u)} du - \binom{k}{2} \int_{t'}^{t'+t} \frac{1}{\lambda(u)} du} dt' dt -$$

$$\frac{\binom{j}{2}}{\binom{n+1}{2} - \binom{j}{2}} \int_0^\infty \int_0^\infty t^2 \frac{\binom{n+1}{2}}{\lambda(t')} \frac{\binom{k}{2}}{\lambda(t'+t)} e^{-\binom{n+1}{2} \int_0^{t'} \frac{1}{\lambda(u)} du - \binom{k}{2} \int_{t'}^{t'+t} \frac{1}{\lambda(u)} du} dt' dt.$$

$\square$

The following chart illustrates the setup of recursions and the way of inductive reasoning.

Chapter 2. General coalescent trees

$$E(T_2^2)_2$$

$$E(T_3^2)_3 \quad E(T_2^2)_3$$

$$E(T_4^2)_4 \quad E(T_3^2)_4 \quad \searrow \quad E(T_2^2)_4$$

$$\cdots \quad \cdots \quad \searrow \quad \cdots \quad \searrow \quad \cdots$$

$$E(T_k^2)_n \quad \searrow$$

$$E(T_k^2)_{n+1}$$

To demonstrate how Lemma 2.4 can be successively applied to eventually obtain Equation 2.3, we derive the special case $E(T_{n-2}^2)_n$. Induction starts with the following two equations, which directly result from the joint density of waiting times.

$$E(T_n^2)_n = \int_0^\infty t_n^2 g_n(t_n) dt_n,$$

$$E(T_{n-1}^2)_n = \int_0^\infty \int_0^\infty t_{n-1}^2 g_{n,n-1}(t_n, t_{n-1}) dt_n dt_{n-1},$$

where $g_n(t_n)$ is defined above, and

$$g_{n,n-1}(t_n, t_{n-1}) = \frac{\binom{n}{2}}{\lambda(t_n)} \frac{\binom{n-1}{2}}{\lambda(t_n + t_{n-1})} \exp\{-\binom{n}{2} \int_0^{t_n} \frac{1}{\lambda(u)} du\} \exp\{-\binom{n-1}{2} \int_{t_n}^{t_n+t_{n-1}} \frac{1}{\lambda(u)} du\}.$$

$$\begin{aligned} E(T_{n-2}^2)_n &= E(E(T_{n-2}^2|T_n)_n) \\ &= \int_0^\infty g_n(t_n) E(T_{n-2}^2|T_n = t_n)_n dt_n \\ &= \int_0^\infty \frac{\binom{n}{2}}{\lambda(t_n)} e^{-\binom{n}{2} \int_0^{t_n} \frac{1}{\lambda(u)} du} \int_0^\infty \int_0^\infty t_{n-2}^2 \frac{\binom{n-1}{2}}{\lambda(t_n + t_{n-1})} \frac{\binom{n-2}{2}}{\lambda(t_n + t_{n-1} + t_{n-2})} \times \\ & \quad e^{-\binom{n-1}{2} \int_{t_n}^{t_n+t_{n-1}} \frac{1}{\lambda(u)} du} e^{-\binom{n-2}{2} \int_{t_n+t_{n-1}}^{t_n+t_{n-1}+t_{n-2}} \frac{1}{\lambda(u)} du} dt_n dt_{n-1} dt_{n-2}. \end{aligned}$$

2.1. Variable population size

By making use of Lemma 2.4, we obtain

$$E(T_{n-2}^2)_n = \frac{\binom{n}{2}}{\binom{n}{2}-\binom{n-1}{2}} \int_0^\infty \int_0^\infty t_{n-2}^2 \frac{\binom{n-1}{2}}{\lambda(t_{n-1})} \frac{\binom{n-2}{2}}{\lambda(t_{n-1}+t_{n-2})} e^{-\binom{n-1}{2}\int_0^{t_{n-1}} \frac{1}{\lambda(u)}du} e^{-\binom{n-2}{2}\int_{t_{n-1}}^{t_{n-1}+t_{n-2}} \frac{1}{\lambda(u)}du} dt_{n-1} dt_{n-2}$$

$$- \frac{\binom{n-1}{2}}{\binom{n}{2}-\binom{n-1}{2}} \int_0^\infty \int_0^\infty t_{n-2}^2 \frac{\binom{n}{2}}{\lambda(t_{n-1})} \frac{\binom{n-2}{2}}{\lambda(t_{n-1}+t_{n-2})} e^{-\binom{n}{2}\int_0^{t_{n-1}} \frac{1}{\lambda(u)}du} e^{-\binom{n-2}{2}\int_{t_{n-1}}^{t_{n-1}+t_{n-2}} \frac{1}{\lambda(u)}du} dt_{n-1} dt_{n-2}$$

$$= \sum_{j=n-1}^{n} (-1)^{j+n-1} \frac{\binom{2n-j-1}{2}}{\binom{n}{2}-\binom{n-1}{2}} \int_0^\infty \int_0^\infty t_{n-2}^2 \frac{\binom{j}{2}}{\lambda(t_{n-1})} \frac{\binom{n-2}{2}}{\lambda(t_{n-1}+t_{n-2})} e^{-\binom{j}{2}\int_0^{t_{n-1}} \frac{1}{\lambda(u)}du} e^{-\binom{n-2}{2}\int_{t_{n-1}}^{t_{n-1}+t_{n-2}} \frac{1}{\lambda(u)}du} dt_{n-1} dt_{n-2}.$$

In general, it holds that

$$E(T_k^2) = \begin{cases} \sum_{j=k+1}^{n} (-1)^{j+k+1} \frac{\binom{k+1}{2}}{\binom{j}{2}} \alpha_{n,j,k+1} \int_0^\infty \int_0^\infty t^2 \, g_{j,k}(t',t) dt' dt, & 2 \leq k \leq n-1, \\ \int_0^\infty t^2 \, g_n(t) dt, & k = n, \end{cases} \quad (2.3)$$

where

$$g_{j,k}(t',t) = \frac{\binom{j}{2}\binom{k}{2}}{\lambda(t')\lambda(t'+t)} \exp\{-\binom{j}{2} \int_0^{t'} \frac{1}{\lambda(u)} du\} \exp\{-\binom{k}{2} \int_{t'}^{t'+t} \frac{1}{\lambda(u)} du\},$$

and e.g., $g_{3,2}(t_3, t_2)$ is the joint density of $(T_3, T_2)_3$.

Proof of Equation 2.3.

We write

$$E(T_k^2)_{n+1} = E(E(T_k^2|T_{n+1})_{n+1}) = \int_0^\infty g_{n+1}(t_{n+1}) E(T_k^2|T_{n+1} = t_{n+1})_{n+1} dt_{n+1}$$

$$= \int_0^\infty \frac{\binom{n+1}{2}}{\lambda(t_{n+1})} \exp\{-\binom{n+1}{2} \int_0^{t_{n+1}} \frac{1}{\lambda(u)} du\} E(T_k^2)_n^{\lambda_s} dt_{n+1},$$

where

Chapter 2. General coalescent trees

$$E(T_k^2)_n^{\lambda_s} = \sum_{j=k+1}^{n} (-1)^{j+k+1} \frac{\binom{k+1}{2}}{\binom{j}{2}} \alpha_{n,j,k+1} \int_0^\infty \int_0^\infty t^2 \frac{\binom{j}{2}}{\lambda(t_{n+1}+t')} \frac{\binom{k}{2}}{\lambda(t_{n+1}+t'+t)} \times$$

$$\exp\{-\binom{j}{2} \int_{t_{n+1}}^{t_{n+1}+t'} \frac{1}{\lambda(u)} du\} \exp\{-\binom{k}{2} \int_{t_{n+1}+t'}^{t_{n+1}+t'+t} \frac{1}{\lambda(u)} du\} dt' dt,$$

for $2 \leq k \leq n-1$, follows from the induction assumption in analogy to the proof of Equation 2.2. Thus,

$$E(T_k^2)_{n+1} = \int_0^\infty \frac{\binom{n+1}{2}}{\lambda(t_{n+1})} e^{-\binom{n+1}{2} \int_0^{t_{n+1}} \frac{1}{\lambda(u)} du} \left(\sum_{j=k+1}^n (-1)^{j+k+1} \frac{\binom{k+1}{2}}{\binom{j}{2}} \alpha_{n,j,k+1} \times \right.$$

$$\left. \int_0^\infty \int_0^\infty t^2 \frac{\binom{j}{2}}{\lambda(t_{n+1}+t')} \frac{\binom{k}{2}}{\lambda(t_{n+1}+t'+t)} e^{-\binom{j}{2} \int_{t_{n+1}}^{t_{n+1}+t'} \frac{1}{\lambda(u)} du \, -\binom{k}{2} \int_{t_{n+1}+t'}^{t_{n+1}+t'+t} \frac{1}{\lambda(u)} du} dt' dt \right) dt_{n+1}$$

$$= \sum_{j=k+1}^{n} (-1)^{j+k+1} \frac{\binom{k+1}{2}}{\binom{j}{2}} \alpha_{n,j,k+1} \int_0^\infty \int_0^\infty \int_0^\infty t^2 \frac{\binom{n+1}{2}}{\lambda(t_{n+1})} \frac{\binom{j}{2}}{\lambda(t_{n+1}+t')} \frac{\binom{k}{2}}{\lambda(t_{n+1}+t'+t)} \times$$

$$e^{-\binom{n+1}{2} \int_0^{t_{n+1}} \frac{1}{\lambda(u)} du \, -\binom{j}{2} \int_{t_{n+1}}^{t_{n+1}+t'} \frac{1}{\lambda(u)} du \, -\binom{k}{2} \int_{t_{n+1}+t'}^{t_{n+1}+t'+t} \frac{1}{\lambda(u)} du} dt_{n+1} dt' dt.$$

Applying Lemma 2.4, we find

$$E(T_k^2)_{n+1} = \sum_{j=k+1}^{n} (-1)^{j+k+1} \frac{\binom{k+1}{2}}{\binom{j}{2}} \alpha_{n,j,k+1} \underbrace{\frac{\binom{n+1}{2}}{\binom{n+1}{2} - \binom{j}{2}}}_{= \alpha_{n+1,j,k+1}} \int_0^\infty \int_0^\infty t^2 \frac{\binom{j}{2}}{\lambda(t')} \frac{\binom{k}{2}}{\lambda(t'+t)} e^{-\binom{j}{2} \int_0^{t'} \frac{1}{\lambda(u)} du \, -\binom{k}{2} \int_{t'}^{t'+t} \frac{1}{\lambda(u)} du} dt' dt -$$

$$\sum_{j=k+1}^{n} (-1)^{j+k+1} \frac{\binom{k+1}{2}}{\binom{j}{2}} \alpha_{n,j,k+1} \underbrace{\frac{\binom{j}{2}}{\binom{n+1}{2} - \binom{j}{2}}}_{= \frac{\binom{k+1}{2}}{\binom{n+1}{2}} \alpha_{n+1,j,k+1}} \int_0^\infty \int_0^\infty t^2 \frac{\binom{j}{2}}{\lambda(t')} \frac{\binom{k}{2}}{\lambda(t'+t)} e^{-\binom{n+1}{2} \int_0^{t'} \frac{1}{\lambda(u)} du \, -\binom{k}{2} \int_{t'}^{t'+t} \frac{1}{\lambda(u)} du} dt' dt$$

$$= \sum_{j=k+1}^{n} (-1)^{j+k+1} \frac{\binom{k+1}{2}}{\binom{j}{2}} \alpha_{n+1,j,k+1} \int_0^\infty \int_0^\infty t^2 \frac{\binom{j}{2}}{\lambda(t')} \frac{\binom{k}{2}}{\lambda(t'+t)} e^{-\binom{j}{2} \int_0^{t'} \frac{1}{\lambda(u)} du \, -\binom{k}{2} \int_{t'}^{t'+t} \frac{1}{\lambda(u)} du} dt' dt -$$

$$\frac{\binom{k+1}{2}}{\binom{n+1}{2}} \left(\sum_{j=k+1}^{n+1} (-1)^{j+k+1} \alpha_{n+1,j,k+1} - (-1)^{n+1+k+1} \alpha_{n+1,n+1,k+1} \right) \times$$

$$\int_0^\infty \int_0^\infty t^2 \frac{\binom{n+1}{2}}{\lambda(t')} \frac{\binom{k}{2}}{\lambda(t'+t)} e^{-\binom{n+1}{2} \int_0^{t'} \frac{1}{\lambda(u)} du \, -\binom{k}{2} \int_{t'}^{t'+t} \frac{1}{\lambda(u)} du} dt' dt$$

2.1. Variable population size

$= $ (Lemma 2.3)

$$\sum_{j=k+1}^{n}(-1)^{j+k+1}\frac{\binom{k+1}{2}}{\binom{j}{2}}\alpha_{n+1,j,k+1}\int_0^\infty\int_0^\infty t^2\frac{\binom{j}{2}}{\lambda(t')}\frac{\binom{k}{2}}{\lambda(t'+t)}e^{-\binom{j}{2}\int_0^{t'}\frac{1}{\lambda(u)}du\;-\binom{k}{2}\int_{t'}^{t'+t}\frac{1}{\lambda(u)}du}\,dt'dt\;+$$

$$(-1)^{n+1+k+1}\frac{\binom{k+1}{2}}{\binom{n+1}{2}}\alpha_{n+1,n+1,k+1}\int_0^\infty\int_0^\infty t^2\frac{\binom{n+1}{2}}{\lambda(t')}\frac{\binom{k}{2}}{\lambda(t'+t)}e^{-\binom{n+1}{2}\int_0^{t'}\frac{1}{\lambda(u)}du\;-\binom{k}{2}\int_{t'}^{t'+t}\frac{1}{\lambda(u)}du}\,dt'dt$$

$$=\sum_{j=k+1}^{n+1}(-1)^{j+k+1}\frac{\binom{k+1}{2}}{\binom{j}{2}}\alpha_{n+1,j,k+1}\int_0^\infty\int_0^\infty t^2\frac{\binom{j}{2}}{\lambda(t')}\frac{\binom{k}{2}}{\lambda(t'+t)}\exp\{-\binom{j}{2}\int_0^{t'}\frac{1}{\lambda(u)}du\}\exp\{-\binom{k}{2}\int_{t'}^{t'+t}\frac{1}{\lambda(u)}du\}dt'dt$$

$\square$

For all demographic models with a finite number of m intervals, each with a different constant population size, Equation 2.3 can be explicitly solved by decomposition of the double integrals according to the $\binom{m+1}{2}$ possible arrangements of coalescent events for a sample of size 3 over these m time phases. For example, in the case of the previously mentioned bottleneck model (cf. Fig. 2.2) it is instructive to express the solution for 3 alleles, and then to replace $\binom{3}{2}$ by $\binom{j}{2}$ and $\binom{2}{2}$ by $\binom{k}{2}$. In the case of exponential population growth, one must resort to a numerical evaluation of Equation 2.3.

2.1.3 Mean product of two distinct waiting times

The following equation is the quintessence of the recursive procedure for the derivation of the mean product of two distinct waiting times.

Lemma 2.5.

$$\int_0^\infty\int_0^\infty\int_0^\infty t't\frac{\binom{n+1}{2}}{\lambda(t_{n+1})}\frac{\binom{j}{2}}{\lambda(t_{n+1}+t')}\frac{\binom{i}{2}}{\lambda(t_{n+1}+t'+t)}e^{-\binom{n+1}{2}\int_0^{t_{n+1}}\frac{1}{\lambda(u)}du\;-\binom{j}{2}\int_{t_{n+1}}^{t_{n+1}+t'}\frac{1}{\lambda(u)}du\;-\binom{i}{2}\int_{t_{n+1}+t'}^{t_{n+1}+t'+t}\frac{1}{\lambda(u)}du}\,dt_{n+1}dt'dt =$$

$$\frac{\binom{n+1}{2}}{\binom{n+1}{2}-\binom{j}{2}}\int_0^\infty\int_0^\infty t't\frac{\binom{j}{2}}{\lambda(t')}\frac{\binom{i}{2}}{\lambda(t'+t)}e^{-\binom{j}{2}\int_0^{t'}\frac{1}{\lambda(u)}du\;-\binom{i}{2}\int_{t'}^{t'+t}\frac{1}{\lambda(u)}du}\,dt'dt\;-$$

$$\frac{\binom{j}{2}(\binom{n+1}{2}-\binom{i}{2})}{(\binom{j}{2}-\binom{i}{2})(\binom{n+1}{2}-\binom{j}{2})}\int_0^\infty\int_0^\infty t't\frac{\binom{n+1}{2}}{\lambda(t')}\frac{\binom{i}{2}}{\lambda(t'+t)}e^{-\binom{n+1}{2}\int_0^{t'}\frac{1}{\lambda(u)}du\;-\binom{i}{2}\int_{t'}^{t'+t}\frac{1}{\lambda(u)}du}\,dt'dt\;+$$

$$\frac{\binom{j}{2}}{\binom{j}{2}-\binom{i}{2}}\int_0^\infty\int_0^\infty t't\frac{\binom{n+1}{2}}{\lambda(t')}\frac{\binom{j}{2}}{\lambda(t'+t)}e^{-\binom{n+1}{2}\int_0^{t'}\frac{1}{\lambda(u)}du\;-\binom{j}{2}\int_{t'}^{t'+t}\frac{1}{\lambda(u)}du}\,dt'dt.$$

Chapter 2. General coalescent trees

Proof.

$$\int_0^\infty \int_0^\infty \int_0^\infty t' t \frac{\binom{n+1}{2}}{\lambda(t_{n+1})} \frac{\binom{j}{2}}{\lambda(t_{n+1}+t')} \frac{\binom{i}{2}}{\lambda(t_{n+1}+t'+t)} e^{-\binom{n+1}{2}\int_0^{t_{n+1}} \frac{1}{\lambda(u)}du} e^{-\binom{j}{2}\int_{t_{n+1}}^{t_{n+1}+t'} \frac{1}{\lambda(u)}du} e^{-\binom{i}{2}\int_{t_{n+1}+t'}^{t_{n+1}+t'+t} \frac{1}{\lambda(u)}du} dt_{n+1} dt' dt =$$

(substitute $t_{n+1}+t'+t$ by s_2 and $t_{n+1}+t'$ by s_1, respectively)

$$\int_0^\infty \int_0^{s_2} \int_0^{s_1} (s_1 - t_{n+1})(s_2 - s_1) \frac{\binom{n+1}{2}}{\lambda(t_{n+1})} \frac{\binom{j}{2}}{\lambda(s_1)} \frac{\binom{i}{2}}{\lambda(s_2)} e^{-\binom{n+1}{2}\int_0^{t_{n+1}} \frac{1}{\lambda(u)}du} e^{-\binom{j}{2}\int_{t_{n+1}}^{s_1} \frac{1}{\lambda(u)}du} e^{-\binom{i}{2}\int_{s_1}^{s_2} \frac{1}{\lambda(u)}du} dt_{n+1} ds_1 ds_2 =$$

$$\int_0^\infty \int_0^{s_2} \int_0^{s_1} \left(s_1(s_2 - s_1) - t_{n+1}(s_2 - t_{n+1}) + t_{n+1}(s_1 - t_{n+1}) \right) \frac{\binom{n+1}{2}}{\lambda(t_{n+1})} \frac{\binom{j}{2}}{\lambda(s_1)} \frac{\binom{i}{2}}{\lambda(s_2)} \times$$

$$e^{-\binom{n+1}{2}\int_0^{t_{n+1}} \frac{1}{\lambda(u)}du} e^{-\binom{j}{2}\int_{t_{n+1}}^{s_1} \frac{1}{\lambda(u)}du} e^{-\binom{i}{2}\int_{s_1}^{s_2} \frac{1}{\lambda(u)}du} dt_{n+1} ds_1 ds_2 =$$

$$\int_0^\infty \int_0^{s_2} s_1(s_2 - s_1) \frac{\binom{j}{2}}{\lambda(s_1)} \frac{\binom{i}{2}}{\lambda(s_2)} \frac{\binom{n+1}{2}}{\binom{n+1}{2} - \binom{j}{2}} \left(\underbrace{\int_0^{s_1} \frac{\binom{n+1}{2} - \binom{j}{2}}{\lambda(t_{n+1})} e^{-(\binom{n+1}{2} - \binom{j}{2}) \int_0^{t_{n+1}} \frac{1}{\lambda(u)}du} dt_{n+1}}_{= 1 - e^{-(\binom{n+1}{2} - \binom{j}{2}) \int_0^{s_1} \frac{1}{\lambda(u)}du}} \right) \times$$

$$e^{-\binom{j}{2}\int_0^{s_1} \frac{1}{\lambda(u)}du} e^{-\binom{i}{2}\int_{s_1}^{s_2} \frac{1}{\lambda(u)}du} ds_1 ds_2 -$$

$$\int_0^\infty \int_0^{s_2} t_{n+1}(s_2 - t_{n+1}) \frac{\binom{n+1}{2}}{\lambda(t_{n+1})} \frac{\binom{i}{2}}{\lambda(s_2)} \frac{\binom{j}{2}}{\binom{j}{2} - \binom{i}{2}} \left(\underbrace{\int_{t_{n+1}}^{s_2} \frac{\binom{j}{2} - \binom{i}{2}}{\lambda(s_1)} e^{-(\binom{j}{2} - \binom{i}{2})\int_{t_{n+1}}^{s_1} \frac{1}{\lambda(u)}du} ds_1}_{= 1 - e^{-(\binom{j}{2} - \binom{i}{2})\int_{t_{n+1}}^{s_2} \frac{1}{\lambda(u)}du}} \right) \times$$

$$e^{-\binom{n+1}{2}\int_0^{t_{n+1}} \frac{1}{\lambda(u)}du} e^{-\binom{i}{2}\int_{t_{n+1}}^{s_2} \frac{1}{\lambda(u)}du} dt_{n+1} ds_2 +$$

$$\int_0^\infty \int_0^{s_1} t_{n+1}(s_1 - t_{n+1}) \frac{\binom{n+1}{2}}{\lambda(t_{n+1})} \frac{\binom{j}{2}}{\lambda(s_1)} \left(\underbrace{\int_{s_1}^\infty \frac{\binom{i}{2}}{\lambda(s_2)} e^{-\binom{i}{2}\int_{s_1}^{s_2} \frac{1}{\lambda(u)}du} ds_2}_{= 1} \right) \times$$

$$e^{-\binom{n+1}{2}\int_0^{t_{n+1}} \frac{1}{\lambda(u)}du} e^{-\binom{j}{2}\int_{t_{n+1}}^{s_1} \frac{1}{\lambda(u)}du} dt_{n+1} ds_1 =$$

(rename variables)

$$\frac{\binom{n+1}{2}}{\binom{n+1}{2} - \binom{j}{2}} \int_0^\infty \int_0^{s_2} s_1(s_2 - s_1) \frac{\binom{j}{2}}{\lambda(s_1)} \frac{\binom{i}{2}}{\lambda(s_2)} e^{-\binom{j}{2}\int_0^{s_1} \frac{1}{\lambda(u)}du} e^{-\binom{i}{2}\int_{s_1}^{s_2} \frac{1}{\lambda(u)}du} ds_1 ds_2 -$$

$$\frac{\binom{j}{2}}{\binom{n+1}{2} - \binom{j}{2}} \int_0^\infty \int_0^{s_2} s_1(s_2 - s_1) \frac{\binom{n+1}{2}}{\lambda(s_1)} \frac{\binom{i}{2}}{\lambda(s_2)} e^{-\binom{n+1}{2}\int_0^{s_1} \frac{1}{\lambda(u)}du} e^{-\binom{i}{2}\int_{s_1}^{s_2} \frac{1}{\lambda(u)}du} ds_1 ds_2 -$$

2.1. Variable population size

$$\frac{\binom{j}{2}}{\binom{j}{2}-\binom{i}{2}}\int_0^\infty\int_0^{s_2}s_1(s_2-s_1)\frac{\binom{n+1}{2}}{\lambda(s_1)}\frac{\binom{i}{2}}{\lambda(s_2)}e^{-\binom{n+1}{2}\int_0^{s_1}\frac{1}{\lambda(u)}du - \binom{j}{2}\int_{s_1}^{s_2}\frac{1}{\lambda(u)}du}ds_1ds_2 +$$

$$\frac{\binom{i}{2}}{\binom{j}{2}-\binom{i}{2}}\int_0^\infty\int_0^{s_2}s_1(s_2-s_1)\frac{\binom{n+1}{2}}{\lambda(s_1)}\frac{\binom{j}{2}}{\lambda(s_2)}e^{-\binom{n+1}{2}\int_0^{s_1}\frac{1}{\lambda(u)}du - \binom{i}{2}\int_{s_1}^{s_2}\frac{1}{\lambda(u)}du}ds_1ds_2 +$$

$$\int_0^\infty\int_0^{s_2}s_1(s_2-s_1)\frac{\binom{n+1}{2}}{\lambda(s_1)}\frac{\binom{j}{2}}{\lambda(s_2)}e^{-\binom{n+1}{2}\int_0^{s_1}\frac{1}{\lambda(u)}du - \binom{i}{2}\int_{s_1}^{s_2}\frac{1}{\lambda(u)}du}ds_1ds_2 =$$

(introduce $s_2 - s_1 = t$, rename s_1 by t', replace s_2 by $t' + t$)

$$\frac{\binom{n+1}{2}}{\binom{n+1}{2}-\binom{j}{2}}\int_0^\infty\int_0^\infty t't\frac{\binom{j}{2}}{\lambda(t')}\frac{\binom{i}{2}}{\lambda(t'+t)}e^{-\binom{j}{2}\int_0^{t'}\frac{1}{\lambda(u)}du - \binom{i}{2}\int_{t'}^{t'+t}\frac{1}{\lambda(u)}du}dt'dt -$$

$$\frac{\binom{j}{2}(\binom{n+1}{2}-\binom{i}{2})}{(\binom{j}{2}-\binom{i}{2})(\binom{n+1}{2}-\binom{j}{2})}\int_0^\infty\int_0^\infty t't\frac{\binom{n+1}{2}}{\lambda(t')}\frac{\binom{i}{2}}{\lambda(t'+t)}e^{-\binom{n+1}{2}\int_0^{t'}\frac{1}{\lambda(u)}du - \binom{i}{2}\int_{t'}^{t'+t}\frac{1}{\lambda(u)}du}dt'dt +$$

$$\frac{\binom{i}{2}}{\binom{j}{2}-\binom{i}{2}}\int_0^\infty\int_0^\infty t't\frac{\binom{n+1}{2}}{\lambda(t')}\frac{\binom{j}{2}}{\lambda(t'+t)}e^{-\binom{n+1}{2}\int_0^{t'}\frac{1}{\lambda(u)}du - \binom{j}{2}\int_{t'}^{t'+t}\frac{1}{\lambda(u)}du}dt'dt.$$

□

The following chart illustrates the set-up of recursions upon which the proof by induction for Equation 2.4 rests.

$$
\begin{array}{ccccccccccc}
 & & & & & & E(T_3T_2)_3 & & & & \\
 & & & & & & \downarrow & & & & \\
 & & & & E(T_4T_3)_4 & & E(T_4T_2)_4 & \searrow & E(T_3T_2)_4 & & \\
 & & & & \downarrow & & \downarrow & & & & \\
 & & E(T_5T_4)_5 & & E(T_5T_3)_5 & \searrow & E(T_4T_3)_5 & & E(T_5T_2)_5 & \searrow & E(T_4T_2)_5 & \searrow & E(T_3T_2)_5 \\
 & & \downarrow & & \downarrow & & & & \downarrow & & & & \\
\underline{E(T_6T_5)_6} & & E(T_6T_4)_6 & \searrow & E(T_5T_4)_6 & & E(T_6T_3)_6 & \searrow & E(T_5T_3)_6 & \searrow & E(T_4T_3)_6 & E(T_6T_2)_6 & \searrow & E(T_5T_2)_6 & \searrow & E(T_4T_2)_6 & \searrow & E(T_3T_2)_6 \\
\end{array}
$$

The repeated application of Lemma 2.5 eventually leads to

$$E(T_{k'}T_k) = \sum_{\substack{j=k' \\ i\neq j}}^{n}\sum_{i=k}^{k'}(-1)^{i+j+k+k'}\frac{\binom{j}{2}-\binom{i}{2}}{\binom{j}{2}}\alpha_{n,j,k'}\,\alpha_{k',i,k}\int_0^\infty\int_0^\infty t't\,g_{j,i}(t',t)dt'dt, \quad (2.4)$$

for $2 \leq k < k' \leq n$, where

$$g_{j,i}(t',t) = \frac{\binom{j}{2}\binom{i}{2}}{\lambda(t')\lambda(t'+t)}\exp\{-\binom{j}{2}\int_0^{t'}\frac{1}{\lambda(u)}du\}\exp\{-\binom{i}{2}\int_{t'}^{t'+t}\frac{1}{\lambda(u)}du\}.$$

Chapter 2. General coalescent trees

Proof of Equation 2.4.

For the initial step (underlined in the chart) of the proof by induction of Equation 2.4, we have

$$E(T_n T_{n-1}) = \int_0^\infty \int_0^\infty t' t \frac{\binom{n}{2}}{\lambda(t')} \frac{\binom{n-1}{2}}{\lambda(t'+t)} \exp\{-\binom{n}{2} \int_0^{t'} \frac{1}{\lambda(u)} du\} \exp\{-\binom{n-1}{2} \int_{t'}^{t'+t} \frac{1}{\lambda(u)} du\} dt' dt.$$

This follows directly from the joint density of waiting times. As illustrated in the chart above, the induction step separates into two cases. First, we show how to infer $E(T_{n+1}T_k)_{n+1}$ from the induction assumption $E(T_n T_k)_n$ (downward arrows in the chart).

Let $n = k' > k \geq 2$. Then Equation 2.4 reads

$$E(T_n T_k)_n = \sum_{i=k}^{n-1} (-1)^{i+k} \underbrace{\frac{\binom{n}{2} - \binom{i}{2}}{\binom{n}{2}} \alpha_{n,i,k}}_{= \alpha_{n-1,i,k}} \int_0^\infty \int_0^\infty t' t \frac{\binom{n}{2}}{\lambda(t')} \frac{\binom{i}{2}}{\lambda(t'+t)} e^{-\binom{n}{2}\int_0^{t'} \frac{1}{\lambda(u)} du} e^{-\binom{i}{2}\int_{t'}^{t'+t} \frac{1}{\lambda(u)} du} dt' dt.$$

We write

$$E(T_{n+1}T_k)_{n+1} = E(E(T_k|T_{n+1})_{n+1} T_{n+1}) = \int_0^\infty g_{n+1}(t_{n+1}) t_{n+1} E(T_k|T_{n+1} = t_{n+1})_{n+1} dt_{n+1},$$

where

$$E(T_k|T_{n+1} = t_{n+1}) = \sum_{i=k}^n (-1)^{i+k} \alpha_{n,i,k} \int_0^\infty \exp\{-\binom{i}{2} \int_{t_{n+1}}^{t_{n+1}+t} \frac{1}{\lambda(u)} du\} dt$$

has been shown in the proof of Equation 2.2.

Therefore,

$$E(T_{n+1}T_k)_{n+1} = \int_0^\infty \frac{\binom{n+1}{2}}{\lambda(t_{n+1})} e^{-\binom{n+1}{2}\int_0^{t_{n+1}} \frac{1}{\lambda(u)} du} t_{n+1} \times$$

$$\left(\sum_{i=k}^n (-1)^{i+k} \alpha_{n,i,k} \int_0^\infty e^{-\binom{i}{2}\int_{t_{n+1}}^{t_{n+1}+t} \frac{1}{\lambda(u)} du} dt \right) dt_{n+1}$$

2.1. Variable population size

$$= \text{(Lemma 2.1)} \int_0^\infty \frac{\binom{n+1}{2}}{\lambda(t_{n+1})} e^{-\binom{n+1}{2}\int_0^{t_{n+1}} \frac{1}{\lambda(u)}du} t_{n+1} \times$$

$$\left(\sum_{i=k}^n (-1)^{i+k} \alpha_{n,i,k} \int_0^\infty t \frac{\binom{i}{2}}{\lambda(t_{n+1}+t)} e^{-\binom{i}{2}\int_{t_{n+1}}^{t_{n+1}+t} \frac{1}{\lambda(u)}du} dt \right) dt_{n+1}$$

$= \text{(rename } t_{n+1} \text{ by } t')$

$$\sum_{i=k}^n (-1)^{i+k} \alpha_{n,i,k} \int_0^\infty \int_0^\infty t' t \frac{\binom{n+1}{2}}{\lambda(t')} \frac{\binom{i}{2}}{\lambda(t'+t)} e^{-\binom{n+1}{2}\int_0^{t'} \frac{1}{\lambda(u)}du - \binom{i}{2}\int_{t'}^{t'+t} \frac{1}{\lambda(u)}du} dt' dt,$$

which completes the proof of the first case.

For the conclusion of $E(T_{k'}T_k)_{n+1}$ from the induction assumption $E(T_{k'}T_k)_n$ (diagonal arrows in the chart), we write

$$E(T_{k'}T_k)_{n+1} = E(E(T_{k'}T_k|T_{n+1})_{n+1}) = \int_0^\infty g_{n+1}(t_{n+1}) E(T_{k'}T_k|T_{n+1}=t_{n+1})_{n+1} dt_{n+1}$$

$$= \int_0^\infty \frac{\binom{n+1}{2}}{\lambda(t_{n+1})} \exp\{-\binom{n+1}{2} \int_0^{t_{n+1}} \frac{1}{\lambda(u)} du\} E(T_{k'}T_k)_n^{\lambda_s} dt_{n+1},$$

where

$$E(T_{k'}T_k)_n^{\lambda_s} = \sum_{j=k'}^n \sum_{\substack{i=k \\ i \neq j}}^{k'} (-1)^{i+j+k+k'} \frac{\binom{j}{2} - \binom{i}{2}}{\binom{j}{2}} \alpha_{n,j,k'} \alpha_{k',i,k} \times$$

$$\int_0^\infty \int_0^\infty t' t \frac{\binom{j}{2}}{\lambda(t_{n+1}+t')} \frac{\binom{i}{2}}{\lambda(t_{n+1}+t'+t)} e^{-\binom{j}{2}\int_{t_{n+1}}^{t_{n+1}+t'} \frac{1}{\lambda(u)}du - \binom{i}{2}\int_{t_{n+1}+t'}^{t_{n+1}+t'+t} \frac{1}{\lambda(u)}du} dt' dt$$

follows from the induction assumption in analogy to the proof of Equation 2.2. Hence,

$$E(T_{k'}T_k)_{n+1} = \int_0^\infty \frac{\binom{n+1}{2}}{\lambda(t_{n+1})} e^{-\binom{n+1}{2}\int_0^{t_{n+1}} \frac{1}{\lambda(u)}du} \left(\sum_{j=k'}^n \sum_{\substack{i=k \\ i \neq j}}^{k'} (-1)^{i+j+k+k'} \frac{\binom{j}{2} - \binom{i}{2}}{\binom{j}{2}} \alpha_{n,j,k'} \alpha_{k',i,k} \times \right.$$

$$\left. \int_0^\infty \int_0^\infty t' t \frac{\binom{j}{2}}{\lambda(t_{n+1}+t')} \frac{\binom{i}{2}}{\lambda(t_{n+1}+t'+t)} e^{-\binom{j}{2}\int_{t_{n+1}}^{t_{n+1}+t'} \frac{1}{\lambda(u)}du - \binom{i}{2}\int_{t_{n+1}+t'}^{t_{n+1}+t'+t} \frac{1}{\lambda(u)}du} dt' dt \right) dt_{n+1}$$

Chapter 2. General coalescent trees

$$= \sum_{\substack{j=k' \\ i \neq j}}^{n} \sum_{i=k}^{k'} (-1)^{i+j+k+k'} \frac{\binom{j}{2} - \binom{i}{2}}{\binom{j}{2}} \alpha_{n,j,k'} \, \alpha_{k',i,k} \int_0^\infty \int_0^\infty \int_0^\infty t' t \frac{\binom{n+1}{2}}{\lambda(t_{n+1})} \frac{\binom{j}{2}}{\lambda(t_{n+1}+t')} \frac{\binom{i}{2}}{\lambda(t_{n+1}+t'+t)} \times$$

$$e^{-\binom{n+1}{2} \int_0^{t_{n+1}} \frac{1}{\lambda(u)} du} \; e^{-\binom{j}{2} \int_{t_{n+1}}^{t_{n+1}+t'} \frac{1}{\lambda(u)} du} \; e^{-\binom{i}{2} \int_{t_{n+1}+t'}^{t_{n+1}+t'+t} \frac{1}{\lambda(u)} du} \, dt_{n+1} \, dt' \, dt.$$

By Lemma 2.5, we have

$$E(T_{k'}T_k)_{n+1} = \sum_{\substack{j=k' \\ i \neq j}}^{n} \sum_{i=k}^{k'} (-1)^{i+j+k+k'} \frac{\binom{j}{2} - \binom{i}{2}}{\binom{j}{2}} \underbrace{\alpha_{n,j,k'} \frac{\binom{n+1}{2}}{\binom{n+1}{2} - \binom{j}{2}}}_{= \alpha_{n+1,j,k'}} \alpha_{k',i,k} \times$$

$$\int_0^\infty \int_0^\infty t' t \frac{\binom{j}{2}}{\lambda(t')} \frac{\binom{i}{2}}{\lambda(t'+t)} e^{-\binom{j}{2} \int_0^{t'} \frac{1}{\lambda(u)} du} e^{-\binom{i}{2} \int_{t'}^{t'+t} \frac{1}{\lambda(u)} du} dt' dt -$$

$$\sum_{\substack{j=k' \\ i \neq j}}^{n} \sum_{i=k}^{k'} (-1)^{i+j+k+k'} \frac{\binom{j}{2} - \binom{i}{2}}{\binom{j}{2}} \alpha_{n,j,k'} \, \alpha_{k',i,k} \frac{\binom{j}{2}(\binom{n+1}{2} - \binom{j}{2})}{(\binom{j}{2} - \binom{i}{2})(\binom{n+1}{2} - \binom{j}{2})} \times$$

$$\int_0^\infty \int_0^\infty t' t \frac{\binom{n+1}{2}}{\lambda(t')} \frac{\binom{i}{2}}{\lambda(t'+t)} e^{-\binom{n+1}{2} \int_0^{t'} \frac{1}{\lambda(u)} du} e^{-\binom{i}{2} \int_{t'}^{t'+t} \frac{1}{\lambda(u)} du} dt' dt +$$

$$\sum_{\substack{j=k' \\ i \neq j}}^{n} \sum_{i=k}^{k'} (-1)^{i+j+k+k'} \frac{\binom{j}{2} - \binom{i}{2}}{\binom{j}{2}} \alpha_{n,j,k'} \, \alpha_{k',i,k} \frac{\binom{j}{2}}{\binom{j}{2} - \binom{i}{2}} \times$$

$$\int_0^\infty \int_0^\infty t' t \frac{\binom{n+1}{2}}{\lambda(t')} \frac{\binom{j}{2}}{\lambda(t'+t)} e^{-\binom{n+1}{2} \int_0^{t'} \frac{1}{\lambda(u)} du} e^{-\binom{j}{2} \int_{t'}^{t'+t} \frac{1}{\lambda(u)} du} dt' dt$$

$$= \sum_{\substack{j=k' \\ i \neq j}}^{n} \sum_{i=k}^{k'} (-1)^{i+j+k+k'} \frac{\binom{j}{2} - \binom{i}{2}}{\binom{j}{2}} \alpha_{n+1,j,k'} \, \alpha_{k',i,k} \times$$

$$\int_0^\infty \int_0^\infty t' t \frac{\binom{j}{2}}{\lambda(t')} \frac{\binom{i}{2}}{\lambda(t'+t)} e^{-\binom{j}{2} \int_0^{t'} \frac{1}{\lambda(u)} du} e^{-\binom{i}{2} \int_{t'}^{t'+t} \frac{1}{\lambda(u)} du} dt' dt -$$

$$\sum_{\substack{i=k \\ i \neq j}}^{k'} (-1)^{i+k} \left(\sum_{j=k'}^{n} (-1)^{j+k'} \underbrace{\alpha_{n,j,k'} \frac{\binom{n+1}{2}}{\binom{n+1}{2} - \binom{j}{2}}}_{= \alpha_{n+1,j,k'}} \right) \alpha_{k',i,k} \frac{\binom{n+1}{2} - \binom{i}{2}}{\binom{n+1}{2}} \times$$

$$\int_0^\infty \int_0^\infty t' t \frac{\binom{n+1}{2}}{\lambda(t')} \frac{\binom{i}{2}}{\lambda(t'+t)} e^{-\binom{n+1}{2} \int_0^{t'} \frac{1}{\lambda(u)} du} e^{-\binom{i}{2} \int_{t'}^{t'+t} \frac{1}{\lambda(u)} du} dt' dt +$$

$$\sum_{\substack{j=k' \\ j \neq i}}^{n} (-1)^{j+k'} \alpha_{n,j,k'} \left(\sum_{i=k}^{k'} (-1)^{i+k} \alpha_{k',i,k} \right) \times$$

$$\int_0^\infty \int_0^\infty t' t \frac{\binom{n+1}{2}}{\lambda(t')} \frac{\binom{j}{2}}{\lambda(t'+t)} e^{-\binom{n+1}{2} \int_0^{t'} \frac{1}{\lambda(u)} du} e^{-\binom{j}{2} \int_{t'}^{t'+t} \frac{1}{\lambda(u)} du} dt' dt.$$

2.1. Variable population size

Finally, and by applying Lemma 2.3, we find

$$\begin{aligned}E(T_{k'}T_k)_{n+1} &= \sum_{\substack{j=k' \\ i \neq j}}^{n}\sum_{i=k}^{k'}(-1)^{i+j+k+k'}\frac{\binom{j}{2}-\binom{i}{2}}{\binom{j}{2}}\alpha_{n+1,j,k'}\,\alpha_{k',i,k} \times \\ & \qquad \int_0^\infty\!\!\int_0^\infty t't\frac{\binom{j}{2}}{\lambda(t')}\frac{\binom{i}{2}}{\lambda(t'+t)}e^{-\binom{j}{2}\int_0^{t'}\frac{1}{\lambda(u)}du}e^{-\binom{i}{2}\int_{t'}^{t'+t}\frac{1}{\lambda(u)}du}\,dt'dt\, - \\ & \sum_{\substack{i=k \\ i\neq j}}^{k'}(-1)^{i+k}\underbrace{\left(\sum_{j=k'}^{n+1}(-1)^{j+k'}\alpha_{n+1,j,k'}-(-1)^{n+1+k'}\alpha_{n+1,n+1,k'}\right)}_{=0}\alpha_{k',i,k}\frac{\binom{n+1}{2}-\binom{i}{2}}{\binom{n+1}{2}} \times \\ & \qquad \int_0^\infty\!\!\int_0^\infty t't\frac{\binom{n+1}{2}}{\lambda(t')}\frac{\binom{i}{2}}{\lambda(t'+t)}e^{-\binom{n+1}{2}\int_0^{t'}\frac{1}{\lambda(u)}du}e^{-\binom{i}{2}\int_{t'}^{t'+t}\frac{1}{\lambda(u)}du}\,dt'dt\, + \\ & \sum_{\substack{j=k' \\ j\neq i}}^{n}(-1)^{j+k'}\alpha_{n,j,k'}\underbrace{\left(\sum_{i=k}^{k'}(-1)^{i+k}\alpha_{k',i,k}\right)}_{=0} \times \\ & \qquad \int_0^\infty\!\!\int_0^\infty t't\frac{\binom{n+1}{2}}{\lambda(t')}\frac{\binom{i}{2}}{\lambda(t'+t)}e^{-\binom{n+1}{2}\int_0^{t'}\frac{1}{\lambda(u)}du}e^{-\binom{i}{2}\int_{t'}^{t'+t}\frac{1}{\lambda(u)}du}\,dt'dt \\ &= \sum_{\substack{j=k' \\ i\neq j}}^{n+1}\sum_{i=k}^{k'}(-1)^{i+j+k+k'}\frac{\binom{j}{2}-\binom{i}{2}}{\binom{j}{2}}\alpha_{n+1,j,k'}\,\alpha_{k',i,k} \times \\ & \qquad \int_0^\infty\!\!\int_0^\infty t't\frac{\binom{j}{2}}{\lambda(t')}\frac{\binom{i}{2}}{\lambda(t'+t)}e^{-\binom{j}{2}\int_0^{t'}\frac{1}{\lambda(u)}du}e^{-\binom{i}{2}\int_{t'}^{t'+t}\frac{1}{\lambda(u)}du}\,dt'dt.\end{aligned}$$

$\square$

In analogy to the mean squared waiting times, Equation 2.4 can be explicitly solved for all demographic models with a finite number of intervals, each with a different constant population size, whereas one must resort to a numerical evaluation of Equation 2.4 in the case of exponential population growth.

2.1.4 MEAN AND VARIANCE OF TWO TREE SIZE MEASURES

Two commonly used measures of the size of the genealogy are the time to the most recent common ancestor, $T_{\mathrm{MRCA}} = \sum_{k=2}^{n} T_k$, and the sum of all branch lengths in the coalescent tree, $T_c = \sum_{k=2}^{n} k\,T_k$, for sample size n. Using the linearity of expectation, we obtain

$$E(T_{\mathrm{MRCA}}) = \sum_{k=2}^{n} E(T_k), \tag{2.5}$$

$$E(T_c) = \sum_{k=2}^{n} k\,E(T_k). \tag{2.6}$$

Chapter 2. General coalescent trees

To evaluate the variance of T_{MRCA} and T_c, we require the variance and covariance of waiting times, given by

$$V(T_k) = E(T_k^2) - E(T_k)^2, \qquad (2.7)$$
$$\text{Cov}(T_{k'}, T_k) = E(T_{k'} T_k) - E(T_{k'}) E(T_k). \qquad (2.8)$$

It follows immediately that

$$V(T_{\text{MRCA}}) = \sum_{k=2}^{n} V(T_k) + 2 \sum_{k=2}^{n-1} \sum_{k'=k+1}^{n} \text{Cov}(T_{k'}, T_k), \qquad (2.9)$$

$$V(T_c) = \sum_{k=2}^{n} k^2 V(T_k) + 2 \sum_{k=2}^{n-1} \sum_{k'=k+1}^{n} k\, k'\, \text{Cov}(T_{k'}, T_k). \qquad (2.10)$$

2.2 Measures of DNA polymorphism

2.2.1 The number of segregating sites

Let S_n be the number of segregating sites in a sample of size n under the infinitely-many-sites model. Since mutations occur according to a Poisson process on the edges of the tree, S_n is a compound Poisson-distributed random variable with mean $\theta/2\, T_c$ and it holds (e.g., HUDSON 1990, GRIFFITHS and TAVARÉ 2003) that

$$E(S_n) = \frac{\theta}{2} E(T_c), \qquad (2.11)$$
$$V(S_n) = \frac{\theta}{2} E(T_c) + \frac{\theta^2}{4} V(T_c). \qquad (2.12)$$

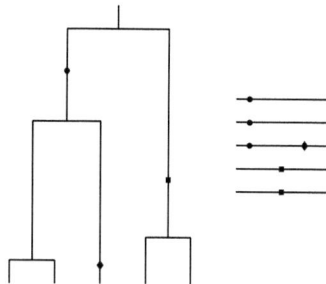

FIGURE 2.3 Mutations of sizes one, two, and three in a coalescent tree for $n = 5$.

2.2. Measures of DNA polymorphism

2.2.2 MUTATIONS OF A CERTAIN SIZE

FU (1995) defined the size of a branch as the number of sequences in a sample that are descendants of that branch (at present time). Furthermore, a mutation is of size i ($1 \leq i \leq n-1$) if it occurs on a branch of size i. Let ξ_i be the number of mutations of size i in a sample of size n. FU (1995) derived expectations, variances and covariances of these random variables for constant population size.

Below, we will follow the article of FU (1995) and extend those results to this paper, which are required to derive the analogous solutions for general coalescent trees. The probabilities in Equations $14-21$ in FU (1995), which rely on a Pólya urn model, are valid in a general coalescent tree since the timing of the urn draws is irrelevant for the binary coloring scheme. The probability $p_{n,k}(i)$ that a randomly chosen line of waiting time T_k is of size i in waiting time T_n (FU 1995, GRIFFITHS and TAVARÉ 1998) is

$$p_{n,k}(i) = \frac{\binom{n-i-1}{k-2}}{\binom{n-1}{k-1}}, \ 1 \leq i \leq n-1. \tag{2.13}$$

The probability $p_{n,k}(i,j)$ that two randomly chosen lines of waiting time T_k are of size i and j, respectively, in waiting time T_n (cf. FU 1995) is

$$p_{n,k}(i,j) = \begin{cases} \frac{\binom{n-i-j-1}{k-3}}{\binom{n-1}{k-1}} = \frac{k-1}{n-k+1} p_{n,k-1}(i+j), & i+j < n \, (\Leftrightarrow 3 \leq k \leq n), \\ \frac{1}{n-1}, & i+j = n \, (\Leftrightarrow k=2). \end{cases} \tag{2.14}$$

Let $2 \leq k < k' \leq n$, without loss of generality. The probability $p_{n,k,k'}(i,j)$ that a randomly chosen line of waiting time T_k and a randomly chosen line of waiting time $T_{k'}$ are of size i and j, respectively, in waiting time T_n is broken down into the two probabilities $p^a_{n,k,k'}(i,j)$ and $p^b_{n,k,k'}(i,j)$ (FU 1995). In the course of subsequent case distinctions for these two probabilities, meaningless combinations of i and j that occur with probability zero are also noted. The probability $p^a_{n,k,k'}(i,j)$ that the line of waiting time $T_{k'}$ is a descendant of the line of waiting time T_k and that these lines are of size j and i, respectively, is given by

$$p^a_{n,k,k'}(i,j) = \begin{cases} \sum_{t=2}^{\min(k'-k+1, i-j+1)} \frac{\binom{k'-k}{t-1}}{\binom{k'-1}{t}} \frac{k-1}{k'} \frac{\binom{i-j-1}{t-2}\binom{n-i-1}{k'-t-1}}{\binom{n-1}{k'-1}}, & i > j, \\ \frac{k-1}{k'(k'-1)} \frac{\binom{n-i-1}{k'-2}}{\binom{n-1}{k'-1}}, & i = j, \\ 0, & \text{otherwise.} \end{cases}$$

Chapter 2. General coalescent trees

The probability $p^b_{n,k,k'}(i,j)$ that the line of waiting time $T_{k'}$ is not a descendant of the line of waiting time T_k and that these lines are of size j and i, respectively, is given by

$$p^b_{n,k,k'}(i,j) = \begin{cases} \sum_{t=1}^{\min(k'-2,k'-k+1,i)} \frac{\binom{k'-k}{t-1}}{\binom{k'-1}{t}} \frac{(k-1)(k'-t)}{tk'} \frac{\binom{i-1}{t-1}\binom{n-i-j-1}{k'-t-2}}{\binom{n-1}{k'-1}}, & i+j < n, \\ \frac{1}{k'(k'-1)} \frac{\binom{n-j-1}{k'-2}}{\binom{n-1}{k'-1}}, & i+j = n,\ k=2, \\ 0, & \text{otherwise.} \end{cases}$$

Therefore,

$$p_{n,k,k'}(i,j) = p^a_{n,k,k'}(i,j) + p^b_{n,k,k'}(i,j).$$

The probabilities $p^a_{n,k,k'}(i,j)$ for $i > j$ and $p^b_{n,k,k'}(i,j)$ for $i+j < n$ given in the above equations include the summation over an additional variable t, which is the size of a randomly chosen line of waiting time T_k in waiting time $T_{k'}$, and will be resolved in the following. For notational convenience, we define

$$p^*_{n,k,k'}(i,j) = \binom{k'-1}{k-1} p_{n,k}(i)\, p_{n-k+1,k'-k+1}(j). \tag{2.15}$$

First, we rewrite the probability $p^a_{n,k,k'}(i,j)$, for $i > j$, as

$$\begin{aligned} p^a_{n,k,k'}(i,j) &= \frac{\binom{n-i-1}{k-2}\binom{n-k-j}{k'-k-1}}{k'\binom{n-1}{k'-1}\binom{k'-1}{k-1}} \sum_{t\geq 0}(t+2)\frac{\binom{i-j-1}{t}\binom{n-k-j-(i-j-1)}{k'-k-1-t}}{\binom{n-k-j}{k'-k-1}} \\ &= \frac{\binom{n-i-1}{k-2}\binom{n-k-j}{k'-k-1}}{k'\binom{n-1}{k'-1}\binom{k'-1}{k-1}} \sum_{t\geq 0}(t+2)p_t, \end{aligned}$$

where p_t are the probabilities of the hypergeometric distribution $H_{k'-k-1;n-k-j,i-j-1}$.

Hence, for $i > j$, we have

$$p^a_{n,k,k'}(i,j) = \begin{cases} \frac{\binom{n-i-1}{k-2}\binom{n-k-j}{k'-k-1}}{k'\binom{n-1}{k'-1}\binom{k'-1}{k-1}}\left((k'-k-1)\frac{i-j-1}{n-k-j}+2\right), & j < n-k, \\ \left(\binom{k+1}{2}\binom{n-1}{k}\right)^{-1}, & j = n-k = n-k'+1, \\ & i = j+1, \\ 0, & \text{otherwise.} \end{cases}$$

2.2. Measures of DNA polymorphism

From this equation it is simple to see that

$$p^a_{n,k,k'}(i,j) = \begin{cases} \sum_{u=1}^{2}(-1)^u \frac{k-1}{k'\binom{k'-1}{k-u}} p^*_{n,k-u+2,k'}(i,j), & i > j, \\ \frac{k-1}{k'(k'-1)} p_{n,k'}(i), & i = j, \\ 0, & \text{otherwise.} \end{cases} \qquad (2.16)$$

The probability $p^b_{n,k,k'}(i,j)$, for $i+j < n$, can be rewritten as

$$\begin{aligned} p^b_{n,k,k'}(i,j) &= \frac{\binom{n-i-j}{k-2}\binom{n-k-j+1}{k'-k}}{(n-i-j)k'\binom{n-1}{k-1}\binom{n-k}{k'-k}} \times \\ &\quad \sum_{t \geq 0}(t^2 - (2k'-3)t + (k'(k'-3)+2))\frac{\binom{i-1}{t}\binom{n-k-j+1-(i-1)}{k'-k-t}}{\binom{n-k-j+1}{k'-k}} \\ &= \frac{\binom{n-i-j}{k-2}\binom{n-k-j+1}{k'-k}}{(n-i-j)k'\binom{n-1}{k-1}\binom{n-k}{k'-k}} \sum_{t \geq 0}(t^2 - (2k'-3)t + (k'(k'-3)+2))p_t, \end{aligned}$$

where p_t are the probabilities of the hypergeometric distribution $H_{k'-k;n-k-j+1,i-1}$.

Therefore, for $i+j < n$, we obtain

$$p^b_{n,k,k'}(i,j) = \begin{cases} \frac{\binom{n-i-j}{k-2}\binom{n-k-j+1}{k'-k}}{(n-i-j)k'\binom{n-1}{k-1}\binom{n-k}{k'-k}} \times \\ \left(\frac{(i-1)(k'-k)((i-1)((k'-k)-1)-(k'-k)+(n-k-j+1))}{(n-k-j+1)(n-k-j)} - \right. \\ \left. (2k'-3)\frac{(k'-k)(i-1)}{n-k-j+1} + (k'(k'-3)+2) \right), & j < n-k, \\ \frac{(k-1)(k-(k-1)(i-1))}{(k+1)k\binom{n-1}{k}}, & j = n-k = n-k'+1, \\ & i = 1,2, \\ 0, & \text{otherwise.} \end{cases}$$

Chapter 2. General coalescent trees

Then it is straightforward to derive

$$p^b_{n,k,k'}(i,j) = \begin{cases} \frac{k-1}{k'\binom{k'-1}{k-1}} \sum_{u=1}^{2} \sum_{v=1}^{2} p^*_{n,k-u+v,k'}(i+j,j), & i+j < n, \\ \frac{1}{k'(k'-1)} p_{n,k'}(j), & i+j = n,\ k=2, \\ 0, & \text{otherwise.} \end{cases} \quad (2.17)$$

Obviously, also the assembled probability

$$p_{n,k,k'}(i,j) = p^a_{n,k,k'}(i,j) + p^b_{n,k,k'}(i,j) \quad (2.18)$$

can be expressed in terms of $p_{n,k'}(\cdot)$ and $p^*_{n,\cdot,k'}(\cdot,j)$.

Let ξ_{kl} be the number of mutations which occurred in the lth line of state k and $\varepsilon_{kl}(i)$ be an index variable, such that $\varepsilon_{kl}(i)$ equals one if the lth line of state k is of size i and zero otherwise. Then (FU 1995),

$$\xi_i = \sum_{k=2}^{n} \sum_{l=1}^{k} \varepsilon_{kl}(i) \xi_{kl}.$$

Since the moments of ξ_{kl} are the prerequisite for the moments of ξ_i, we begin with the derivation of the first- and second-order moments of these random variables. For this purpose, we make repeated use of the assumption that mutations occur according to independent Poisson processes of rate $\theta/2$ along the edges of the tree, conditional on the edge lengths of the tree.

(i) We write

$$E(\xi_{kl}) = E(E(\xi_{kl}|T_k)) = \int_0^\infty g_{T_k}(t_k) E(\xi_{kl}|T_k = t_k) dt_k,$$

where g_{T_k} denotes the marginal density of the random variable T_k. Since

$$E(\xi_{kl}|T_k = t_k) = \sum_{j \geq 0} j \frac{(\frac{\theta}{2} t_k)^j}{j!} \exp\{-\frac{\theta}{2} t_k\} = \frac{\theta}{2} t_k,$$

we have

$$E(\xi_{kl}) = \frac{\theta}{2} \int_0^\infty t_k\, g_{T_k}(t_k) dt_k = \frac{\theta}{2} E(T_k). \quad (2.19)$$

2.2. Measures of DNA polymorphism

(ii) In analogy,

$$E(\xi_{kl}^2) = E(E(\xi_{kl}^2|T_k)) = \int_0^\infty g_{T_k}(t_k)E(\xi_{kl}^2|T_k = t_k)dt_k,$$

where

$$E(\xi_{kl}^2|T_k = t_k) = \sum_{j\geq 0} j^2 \frac{(\frac{\theta}{2}t_k)^j}{j!} \exp\{-\frac{\theta}{2}t_k\} = \frac{\theta}{2}t_k + \frac{\theta^2}{4}t_k^2.$$

Therefore,

$$E(\xi_{kl}^2) = \frac{\theta}{2}\int_0^\infty t_k\, g_{T_k}(t_k)dt_k + \frac{\theta^2}{4}\int_0^\infty t_k^2\, g_{T_k}(t_k)dt_k = \frac{\theta}{2}E(T_k) + \frac{\theta^2}{4}E(T_k^2). \qquad (2.20)$$

(iii) We write,

$$E(\xi_{kl}\xi_{kl'}) = E(E(\xi_{kl}\xi_{kl'}|T_k)) = \int_0^\infty g_{T_k}(t_k)E(\xi_{kl}\xi_{kl'}|T_k = t_k)dt_k,$$

where

$$E(\xi_{kl}\xi_{kl'}|T_k = t_k) = E(\xi_{kl}|T_k = t_k)E(\xi_{kl'}|T_k = t_k) = \frac{\theta^2}{4}t_k^2,$$

due to the conditional independence of ξ_{kl} and $\xi_{kl'}$. Thus,

$$E(\xi_{kl}\xi_{kl'}) = \frac{\theta^2}{4}\int_0^\infty t_k^2\, g_{T_k}(t_k)dt_k = \frac{\theta^2}{4}E(T_k^2). \qquad (2.21)$$

(iv) Finally,

$$\begin{aligned}E(\xi_{k'l'}\xi_{kl}) &= E(E(\xi_{k'l'}\xi_{kl}|T_{k'},T_k)) \\ &= \int_0^\infty\int_0^\infty g_{T_{k'},T_k}(t_{k'},t_k)E(\xi_{k'l'}\xi_{kl}|T_{k'} = t_{k'}, T_k = t_k)dt_{k'}dt_k,\end{aligned}$$

where $g_{T_{k'},T_k}$ denotes the 2-dimensional marginal density of $(T_{k'}, T_k)$. Since the random variables $\xi_{k'l'}$ and ξ_{kl} are conditionally independent, we have

$$E(\xi_{k'l'}\xi_{kl}|T_{k'} = t_{k'}, T_k = t_k) = E(\xi_{k'l'}|T_{k'} = t_{k'})E(\xi_{kl}|T_k = t_k) = \frac{\theta^2}{4}t_{k'}t_k,$$

Chapter 2. General coalescent trees

and immediately obtain

$$E(\xi_{k'l'}\xi_{kl}) = \frac{\theta^2}{4}\int_0^\infty\int_0^\infty t_{k'}t_k\, g_{T_{k'},T_k}(t_{k'},t_k)dt_{k'}dt_k = \frac{\theta^2}{4}E(T_{k'}T_k). \qquad (2.22)$$

Following Equation 22 of Fu (1995), we find

$$E(\xi_i) = \sum_{k=2}^n \sum_{l=1}^k P(\varepsilon_{kl}(i)=1)E(\xi_{kl}) = \sum_{k=2}^n k\, p_{n,k}(i)E(\xi_{kl}).$$

Applying Equation 2.19 results in

$$E(\xi_i) = \frac{\theta}{2}\sum_{k=2}^n k\, p_{n,k}(i)\, E(T_k). \qquad (2.23)$$

This formula already appears in Griffiths and Tavaré (1998).

Under variable population size and by applying Equation 2.2 to a demographic history, which consists of multiple epochs of different but constant population sizes, Equation 2.23 results in Equation 1 of Marth et al. (2004).

To obtain $V(\xi_i)$ and $\mathrm{Cov}(\xi_i,\xi_j)$, we follow Equation 23 in Fu (1995), given by

$$\begin{aligned}
E(\xi_i\xi_j) &= \sum_{k=2}^n\sum_{k'=2}^n\sum_{l=1}^k\sum_{l'=1}^{k'} P(\varepsilon_{kl}(i)\varepsilon_{k'l'}(j)=1)E(\xi_{k'l'}\xi_{kl})\\
&= \delta_{\{i=j\}}\sum_{k=2}^n k p_{n,k}(i)E(\xi_{kl}^2) + \sum_{k=2}^n k(k-1)p_{n,k}(i,j)E(\xi_{kl}\xi_{kl'}) +\\
&\quad \sum_{k=2}^{n-1}\sum_{k'=k+1}^n kk'\Big(p_{n,k,k'}(i,j)+p_{n,k,k'}(j,i)\Big)E(\xi_{k'l'}\xi_{kl}),
\end{aligned}$$

where $\delta_{\{i=j\}}$ equals one when $i=j$ and zero otherwise. Using Equations 2.13 – 2.22, then by separating cases and after some algebra, one eventually obtains

$$V(\xi_i) = \frac{\theta}{2}\sigma_i + \frac{\theta^2}{4}\sigma_{ii}, \qquad (2.24)$$

$$\mathrm{Cov}(\xi_i,\xi_j) = \frac{\theta^2}{4}\sigma_{ij}, \qquad (2.25)$$

where

$$\sigma_i = \sum_{k=2}^n k\, p_{n,k}(i)\, E(T_k),$$

2.2. Measures of DNA polymorphism

and

$$\sigma_{ii} = \begin{cases} \sigma_{ii1} + \sigma_{ii2} - \sigma_i^2, & i < \frac{n}{2}, \\ \sigma_{ii1} + \sigma_{ii3} - \sigma_i^2, & i = \frac{n}{2}, \\ \sigma_{ii1} - \sigma_i^2, & i > \frac{n}{2}, \end{cases}$$

where

$$\sigma_{ii1} = \sum_{k=2}^{n} k p_{n,k}(i) E(T_k^2) + 2 \sum_{k=2}^{n-1} \sum_{k'=k+1}^{n} \frac{k(k-1)}{k'-1} p_{n,k'}(i) E(T_{k'} T_k),$$

$$\sigma_{ii2} = \sum_{k=2}^{n} \frac{k(k-1)^2}{n-k+1} p_{n,k-1}(2i) E(T_k^2) +$$

$$2 \sum_{k=2}^{n-1} \sum_{k'=k+1}^{n} \frac{k(k-1)}{\binom{k'-1}{k-1}} \sum_{u=1}^{2} \sum_{v=1}^{2} p_{n,k-u+v,k'}^{*}(2i,i) E(T_{k'} T_k),$$

$$\sigma_{ii3} = \frac{2}{n-1} E(T_2^2) + \sum_{k'=3}^{n} \frac{4}{k'-1} p_{n,k'}(i) E(T_{k'} T_2).$$

For $i > j$,

$$\sigma_{ij} = \begin{cases} \sigma_{ij1} + \sigma_{ij2} - \sigma_i \sigma_j, & i + j < n, \\ \sigma_{ij1} + \sigma_{ij3} - \sigma_i \sigma_j, & i + j = n, \\ \sigma_{ij1} - \sigma_i \sigma_j, & i + j > n, \end{cases}$$

where

$$\sigma_{ij1} = \sum_{k=2}^{n-1} \sum_{k'=k+1}^{n} \sum_{u=1}^{2} (-1)^u \frac{k(k-1)}{\binom{k'-1}{k-u}} p_{n,k-u+2,k'}^{*}(i,j) E(T_{k'} T_k),$$

$$\sigma_{ij2} = \sum_{k=3}^{n} \frac{k(k-1)^2}{n-k+1} p_{n,k-1}(i+j) E(T_k^2) +$$

$$\sum_{k=2}^{n-1} \sum_{k'=k+1}^{n} \frac{k(k-1)}{\binom{k'-1}{k-1}} \sum_{u=1}^{2} \sum_{v=1}^{2} (p_{n,k-u+v,k'}^{*}(i+j,i) + p_{n,k-u+v,k'}^{*}(i+j,j)) E(T_{k'} T_k),$$

$$\sigma_{ij3} = \frac{2}{n-1} E(T_2^2) + \sum_{k'=3}^{n} \frac{2}{k'-1} (p_{n,k'}(i) + p_{n,k'}(j)) E(T_{k'} T_2).$$

To decide whether a mutation is of size i, knowledge of the ancestral and derived alleles is required. Without this knowledge, one can only consider the type of a mutation (FU 1995). A mutation is of type i, if it is of size i or $n - i$. If η_i is the number of mutations of type i, then analogous results for $E(\eta_i)$, $V(\eta_i)$ and $\text{Cov}(\eta_i, \eta_j)$ can be derived by using $\eta_i = (\xi_i + \xi_{n-i})(1 + \delta_{\{i=n-i\}})^{-1}$. Since $S_n = \sum_{i=1}^{n-1} \xi_i$, the covariances $\text{Cov}(\xi_i, S_n)$ and $\text{Cov}(\eta_i, S_n)$ can be written in terms of the above expressions.

Chapter 2. General coalescent trees

2.2.3 THE AVERAGE NUMBER OF PAIRWISE DIFFERENCES

Let Π_n be the average number of pairwise differences in the $\binom{n}{2}$ comparisons in a sample of size n. According to FU (1995), we write

$$\Pi_n = \frac{1}{\binom{n}{2}} \sum_{i=1}^{n-1} i(n-i)\xi_i,$$

since the ξ_i's are counted in $i(n-i)$ comparisons between the i descendent and $n-i$ non-descendent sequences of the corresponding branches. The results derived in Chapter 2.2.2 can be used to express moments of Π_n in terms of moments of T_k. After some algebra one eventually obtains

$$E(\Pi_n) = \frac{\theta}{2} \sum_{k=2}^{n} a_{n,k} E(T_k), \qquad (2.26)$$

$$V(\Pi_n) = \frac{\theta}{2} \sum_{k=2}^{n} a_{n,k} c_{n,k} E(T_k) + \frac{\theta^2}{4} \sum_{k=2}^{n} a_{n,k}^2 (V(T_k) + b_{n,k} E(T_k^2)) +$$

$$\frac{\theta^2}{2} \sum_{k=2}^{n-1} \sum_{k'=k+1}^{n} a_{n,k} a_{n,k'} (\operatorname{Cov}(T_{k'}, T_k) + b_{n,k'} E(T_{k'} T_k)), \qquad (2.27)$$

$$\operatorname{Cov}(S_n, \Pi_n) = \frac{\theta}{2} \sum_{k=2}^{n} a_{n,k} E(T_k) + \frac{\theta^2}{4} \sum_{k=2}^{n} k a_{n,k} V(T_k) +$$

$$\frac{\theta^2}{4} \sum_{k=2}^{n-1} \sum_{k'=k+1}^{n} (k a_{n,k'} + k' a_{n,k}) \operatorname{Cov}(T_{k'}, T_k), \qquad (2.28)$$

where

$$a_{n,k} = \frac{2(n+1)(k-1)}{(n-1)(k+1)},$$

$$b_{n,k} = \frac{4(n-k)(n-k-1)}{(n+1)n(k-1)(k+2)(k+3)},$$

$$c_{n,k} = \frac{4k(n^2+1) - 2(n+1)(k-2)(k-3)}{n(n-1)(k+2)(k+3)}.$$

Equation 2.26 has been already derived in GRIFFITHS and TAVARÉ (2003). Under variable population size Equation 2.26 simplifies to $E(\Pi_n) = \theta E(T_2)_2$.

Chapter 3

TESTING NEUTRALITY UNDER VARIABLE POPULATION SIZE

3.1 THE ESTIMATED DEMOGRAPHIC HISTORY OF Drosophila melanogaster

LI and STEPHAN (2006) developed a maximum likelihood method to infer demographic changes and to detect recent selective sweeps in populations of varying size from DNA polymorphism data. Based on a sample of average size $\bar{n} = 12$ that comprises 262 loci of the X-chromosome from an African population of *D. melanogaster* and due to an overall excess of rare derived variants, the authors suggested that an instantaneous population expansion model explains the observed polymorphisms significantly better than the standard neutral model of constant population size. The homologous sequences of *D. simulans*—available for 258 loci—were used as the outgroup to infer the ancestral status of a polymorphic site and to estimate divergence between *D. melanogaster* and *D. simulans*. Under the expansion model, the authors estimated $\hat{\theta}_{A0} = 4N_{A0}\bar{\mu} = 0.0499$, where N_{A0} indicates the current effective population size for the X-chromosome in the African population. Since the *D. melanogaster* lineage split from *D. simulans* approximately 2.3 million years ago (LI *et al.* 1999), the average mutation rate per site per generation was estimated as $\bar{\mu} = 1.45 \times 10^{-9}$, assuming ten generations per year. Then, $\hat{N}_{A0} = 8.603 \times 10^6$ and the estimated time of the expansion is $60,000$ years before the present (corresponding to 0.035 in units of $2\hat{N}_{A0}$ generations). The population size before the expansion was estimated as $0.2 \times \hat{N}_{A0}$. It is important to note that the above and the subsequent analysis for the European population assumed that all SNPs evolve neutrally. Since part of the excess of rare variants may be due to purifying selection, the authors repeated the analysis disregarding the singletons, as suggested by FU (1997) and SMITH and EYRE-WALKER (2002).

Chapter 3. Testing neutrality under variable population size

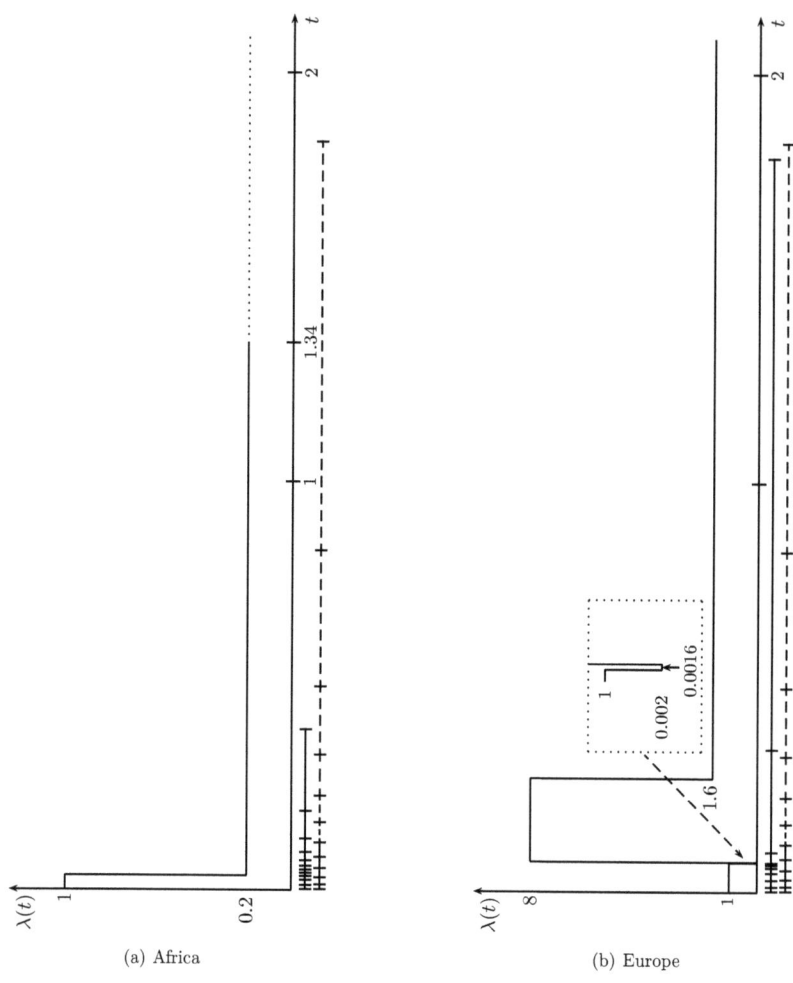

(a) Africa (b) Europe

FIGURE 3.1 The suggested demographic history of the African and European populations of *D. melanogaster* studied by LI and STEPHAN (2006). x-axis: time measured in units of $2N_{A0}$ and $2N_{E0}$ generations, respectively. y-axis: population size in units of N_{A0} and N_{E0}. Ticks below the x-axis indicate expected waiting times $E(T_{12}), \ldots, E(T_2)$ for the bottleneck (solid) and constant population size (dashed lines) models. For the African population all waiting times are heavily and uniformly reduced. The split from *D. simulans* occurs at 1.34 units in the past. For the European population the genealogy consists of short branches in the recent past but elongated branches before the split from the African population. Note that there is only a slight reduction in total tree height.

3.1. The estimated demographic history of *Drosophila melanogaster*

In this case the population size before the expansion was estimated as approximately 0.15, whereas the time of expansion in the past is about 56,000 years and 0.033 in units of $2\hat{N}_{A0}$, respectively.

Furthermore, LI and STEPHAN (2006) studied a sample ($\bar{n} = 12$) from a European population of *D. melanogaster* that consists of 272 X-linked loci, from which 23 loci are non-polymorphic and 6 loci lack the outgroup sequence. To infer the demographic change in the derived European population, LI and STEPHAN (2006) used the joint frequency spectrum (WAKELEY and HEY 1997) that additionally takes into account, if mutations occur exclusively in the African or in the European population, or in both of them. Furthermore, the authors assumed that the out-of-Africa migration did not affect the genetic polymorphism in the African population, since the size of the founder population is likely to be very small compared to the ancestral African population. Therefore, the authors estimated the demographic scenario of the European sample conditional on the estimated demographic scenario of the African sample. The derived estimate for $\hat{\theta}_{E0}(= 4N_{E0}\bar{\mu})$ is 0.0062, with $\hat{N}_{E0} = 1.075 \times 10^6$ being the present size of the European population. The African and European populations diverged approximately 15,800 years ago (corresponding to 0.0735 in units of $2\hat{N}_{E0}$ generations). Subsequently, the European population went through a bottleneck during which its size was reduced to $0.002 \times \hat{N}_{E0}$ and which lasted about 340 years (corresponding to 0.0016 in units of $2\hat{N}_{E0}$ generations). The estimated demographic history of both populations is illustrated in Figure 3.1.

The observed frequency spectrum, the maximum likelihood estimation (MLE) from LI and STEPHAN (2006) and the theoretical frequency spectrum for both populations, obtained by inserting Equation 2.2 into Equation 2.1, are shown in Figure 3.2. The MLE and the theoretical outcome based on the suggested demographic histories are in good agreement since both rely on expected branch lengths. To obtain the theoretical frequency spectrum for the European population joint polymorphisms were ignored, but, interestingly, this simplification had only a marginal effect, as can be seen in Figure 3.2. The property that the theoretical frequency spectrum is strictly monotonically decreasing with the increasing size of a mutation follows directly from Equation 2.1. Since both populations of Drosophila exhibit an excess of high-frequency derived mutations (cf. Figure 3.2), there is an indication for the presence of positive selection in the data (FAY and WU 2000).

Henceforth, $\theta = 4N_0\mu$, where μ is the mutation rate per locus of length L base pairs per generation. Then, the above estimates of θ become $\hat{\theta}_{A0} = 23.5$ ($\bar{L} = 471$) and $\hat{\theta}_{E0} = 3.11$ ($\bar{L} = 498$). For the African and European datasets the average number of segregating sites per locus is 17.87 and 6.88, respectively. Hence, $\hat{\theta}_W$—the standard neutral version of Equation 3.3 as defined by WATTERSON (1975)—is 5.92 for the African, and 2.28 for the European population. In contrast, the estimates based on Equation 3.3, where $E(T_c)$ is consistent with the suggested demographic histories are 23.6 and 3.07, respectively. These discrepancies reflect the differences in total tree length (cf. Figure 3.1). While the theoretical and experimental outcomes of the earlier introduced measures of DNA polymorphism, S_n and Π_n, experience only minor changes with respect to the means, if the different sample sizes of the data are taken into account, the variances of these statistical quantities are more sensitive in this regard. Consequently, for a comparison of the theoretical single-locus results and the experimental data (cf. Table 3.1), the

43

Chapter 3. Testing neutrality under variable population size

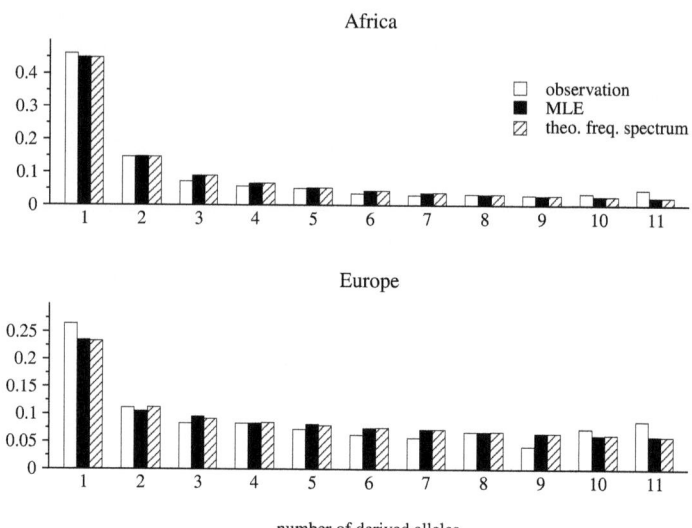

FIGURE 3.2 Frequency spectra of the African and European populations: the open and solid histograms are from LI and STEPHAN (2006). The hatched histogram shows theoretical results for the frequency spectrum.

TABLE 3.1

	Africa*				Europe**		
$\hat{\theta}$	$V(S_{12})$	$E(\Pi_{12})$	$V(\Pi_{12})$	$\hat{\theta}$	$V(S_{12})$	$E(\Pi_{12})$	$V(\Pi_{12})$
			theoretical single-locus results				
	52.18	5.32	7.83		34.57	2.36	6.09
			experimental data				
23.40	91.63	4.93	9.06	2.92	31.11	2.20	4.87

* subset of 204 loci ($n = 12$, $\bar{L} = 474$).
** subset of 235 loci ($n = 12$, $\bar{L} = 498$).
Due to varying locus lengths, the experimental outcomes of S_{12} and Π_{12} are rescaled with respect to $\bar{L}$. Then, $\hat{\theta}$ is estimated from the subsets based on the suggested demographies, and thereafter used for the theoretical results.

subsets of loci with $n = 12$ are extracted from the datasets of both populations. As shown in Table 3.1, there is a pronounced difference with respect to the variance of segregating sites between the theory and the data for the African population. One explanation, besides the possibility that certain loci are exposed to adaptive forces and therefore skew the overall pattern of the putative neutral loci, might be that mutation rates vary across loci. Variations in mutation rate may rather be observable in the African than in the European population because of the about eight-fold larger population size. In the following, the generalized versions of several classical test statistics are analyzed with respect to the estimated demographic scenarios. This is in line with WILLIAMSON et al. (2005), who suggested that it may be more important to find a demographic scenario, which provides a reasonable fit to the data, than the specifics of the demographic model itself. This argument will be explored in more depth in Chapter 4.

3.2 GENERALIZATION OF CLASSICAL TEST STATISTICS

3.2.1 TAJIMA'S D

Tajima's D (TAJIMA 1989b) is a widely applied test statistic for the null hypothesis of neutral evolution under constant population size. In Chapter 2, all necessary ingredients were collected to formulate a generalized version of D, here called D', which may be applied to test the model of neutral evolution under variable population size, conditional on knowledge of the demographic history. The equality

$$\theta = \frac{E(\Pi_n)}{E(T_2)_2} = \frac{2E(S_n)}{E(T_c)}$$

provides the statistical quantity

$$d' = f_{S_n}\Pi_n - f_{\Pi_n}S_n, \tag{3.1}$$

where $f_{S_n} = E(T_c)/2$, $f_{\Pi_n} = E(T_2)_2$. The mean of d' is 0 and the variance of d' is given by

$$V(d') = f_{S_n}^2 V(\Pi_n) + f_{\Pi_n}^2 V(S_n) - 2f_{S_n}f_{\Pi_n}\text{Cov}(S_n, \Pi_n). \tag{3.2}$$

By making use of Equations 2.12, 2.27 and 2.28 together with the unbiased estimators of θ and θ^2, based on S_n, and given by

$$\hat{\theta} = \frac{2S_n}{E(T_c)}, \tag{3.3}$$

$$\hat{\theta^2} = \frac{4S_n(S_n - 1)}{V(T_c) + E(T_c)^2}, \tag{3.4}$$

Chapter 3. Testing neutrality under variable population size

one obtains the test statistic

$$D' = \frac{d'}{\sqrt{\hat{V}(d')}}, \qquad (3.5)$$

where $\hat{V}(d')$ is the estimator of $V(d')$ with respect to Equations 3.3 and 3.4. Clearly, D' simplifies to D for $\lambda(t) = 1$.

Now, we analyze the distributional properties of D' for the African demography, where

$$D' = \frac{0.758\Pi_{12} - 0.228 S_{12}}{\sqrt{(0.0098 + 0.0021 S_{12}) S_{12}}}$$

for $n = 12$. To decide whether an observed value of D' for a certain locus is below or above a critical value, one needs the distribution of D'. In contrast to the standard neutral case, where the distribution of D' can be approximated by a beta distribution (TAJIMA 1989b), a theoretical approximation for the distribution of D' remains unknown under variable population size. Although some intra-locus recombination may occur, recombination is not taken into account. However, the variance of d' under recombination is smaller than in Equation 3.2, such that the critical values of D' obtained without recombination are probably more conservative.

TABLE 3.2 Distributional properties of D' for the African scenario and $n = 12$.

θ	Mean	Variance	*0.05	*0.95
1	−0.010	0.99	−1.22	1.67
3	−0.036	0.96	−1.39	<u>1.69</u>
5	−0.050	0.93	−1.39	1.67
8	−0.067	0.90	−1.47	1.64
10	−0.075	0.88	<u>−1.51</u>	1.60
12	−0.078	0.87	−1.51	1.59
20	−0.091	0.83	−1.49	1.53
40	−0.096	0.80	−1.47	1.48
60	−0.10	0.78	−1.47	1.46

Results are obtained from coalescent simulations with 100,000 replicates.
* 5%- and 95%- quantile, respectively.

3.2. Generalization of classical test statistics

While mutations are assumed to occur at a constant rate per locus, the rates may vary across loci. Therefore, the distribution of D' is generated for different values of θ by coalescent simulations (HUDSON 2002). The two most extreme values of θ—approximately 1 and 60—are estimated by Equation 3.3, which provides conservative—too conservative—values for a possible mutation rate variation range. It may well be that these two extremes simply reflect the random nature of the underlying genealogical trees. This is in line with LI and STEPHAN (2006), who observed that the mutation rate among loci may not be homogeneous but that a varying mutation rate model overestimates the outcome of several summary statistics in the data.

Following TAJIMA (1989b), non-polymorphic iterations are disregarded. Taking the most extreme 5%- and 95%-quantiles among these distributions provides the critical values (underlined in Table 3.2). Simulated distributions of D' for different values of θ and for $n = 12$ and $n = 40$ are shown in Figure 3.3.

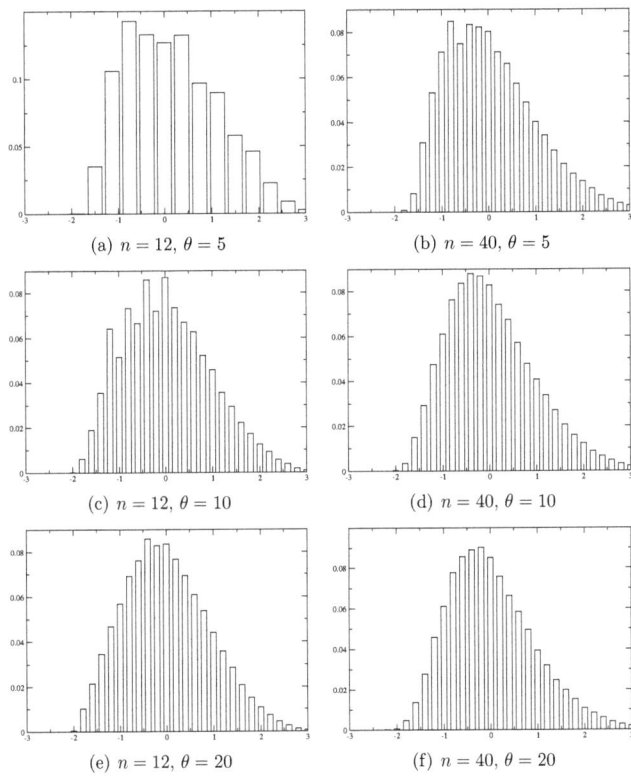

(a) $n = 12, \theta = 5$ (b) $n = 40, \theta = 5$

(c) $n = 12, \theta = 10$ (d) $n = 40, \theta = 10$

(e) $n = 12, \theta = 20$ (f) $n = 40, \theta = 20$

FIGURE 3.3   Simulated distributions of D' for the African scenario.

Chapter 3. Testing neutrality under variable population size

The average of the observed D'-values in the African sample is -0.27, whereas the average of the observed D-values is -0.67. D is smaller than D', since D does not take the effect of population expansion into account. Since not all loci in the African dataset have sample size 12, D' is recalculated for all different sample sizes to obtain the critical values as described above. Based on the critical values, several loci with significantly negative D' (cf. Table 3.3) and one locus with a significantly positive D' are found. Comparing the candidate loci with the results of the likelihood-ratio test (LRT) by LI and STEPHAN (2006), the loci that are marked by an asterisk in Table 3.3 fall into regions for which the null hypothesis of neutral evolution was also rejected in the LRT. Except for Locus 729, all loci are a subset of the 21 significant loci from HUTTER et al. (2007) based on D under constant population size. This decrease in the amount of significant loci is mainly due to our choice of the rejection region by considering a variety of θ−values, whereas the amount of false-positives is rather low ($\sim 10\%$), when applying D under constant population size. However, the critical values obtained by D' are more appropriate to reject the neutral model, since D' takes the suggested demographic history into account.

Now, we analyze the distributional properties of D' for the European demography, where

$$D' = \frac{2.242\Pi_{12} - 0.807 S_{12}}{\sqrt{(0.0485 + 0.0581 S_{12}) S_{12}}}$$

for $n = 12$. The distributions of $d'/\sqrt{V(d')}$ and D' for different values of θ and for sample sizes 12 and 40 are shown in Figure 3.4.

TABLE 3.3 List of outlier loci with respect to D' in the African sample.

locus	25	122	*470	*743	*295	*430	729
D'	-1.60	-1.66	-1.63	-1.70	-1.75	-1.80	2.01
p−value: $\theta = 10$	0.026	0.021	0.019	0.013	0.009	0.006	0.025
$\theta = 20$	0.031	0.025	0.025	0.018	0.014	0.009	0.015
$\theta = 40$	0.032	0.024	0.026	0.018	0.013	0.010	0.012
n	12	12	11	11	12	12	10

Results are obtained from coalescent simulations with 100,000 replicates.

3.2. Generalization of classical test statistics

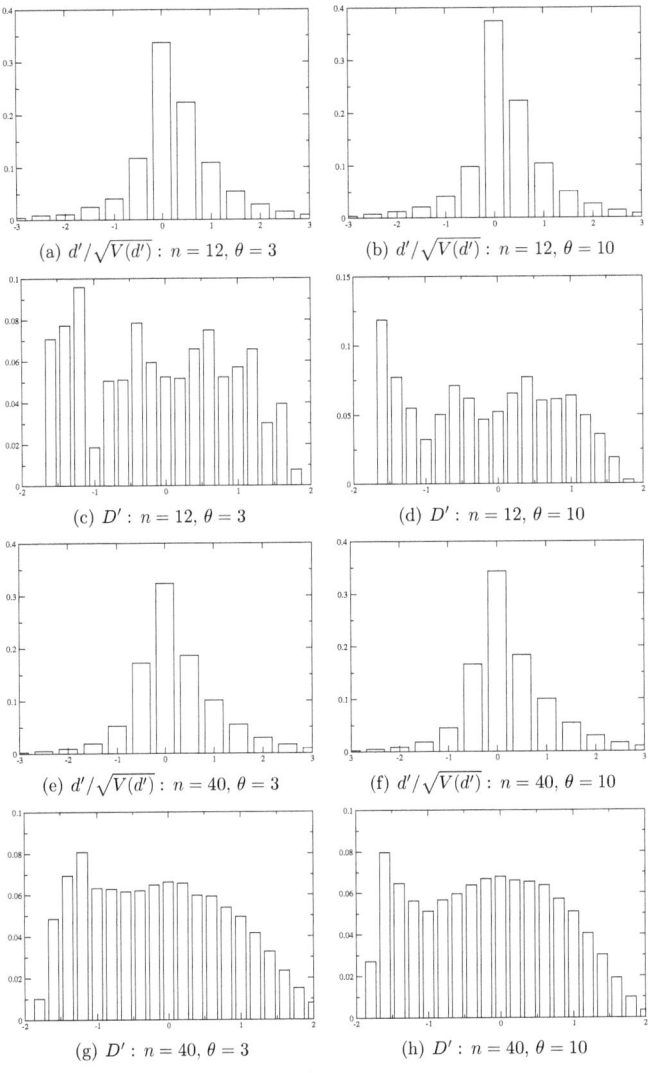

FIGURE 3.4 Simulated distributions of $d'/\sqrt{V(d')}$ and D' for the European scenario.

Chapter 3. Testing neutrality under variable population size

In analogy to TAJIMA (1989b) and FU and LI (1993), the unbiased estimators of θ and θ^2, with respect to S_n, are used, such that the denominator of D' depends solely on the number of segregating sites. Since $d'/\sqrt{V(d')}$ is distributed with mean 0 and variance 1 due to its dependency on θ, the estimators $\hat{\theta}$ and $\hat{\theta}^2$ affect the distortion of the distribution of D' (cf. Figure 3.4). It should be noted that $V(\hat{\theta})$ and $V(\hat{\theta}^2)$ are considerably larger than under constant population size. Furthermore, the distribution of D' for the European scenario is shifted even more towards negative values (($\overline{D'}$, $V(D')$) is about $(-0.2, 0.96)$ and $(-0.3, 0.95)$ for $\theta = 3$ and $\theta = 10$, respectively) under this bottleneck model than under the standard model (cf. Table 1 in TAJIMA 1989b). While the distribution of D becomes smoother with increasing sample size (cf. Figure 3 in TAJIMA 1989b), this is less the case for the distribution of D' and the average θ of about 3 (cf. Figure 3.4). It is interesting to note that, given a demographic scenario such as the European one, increased sample size may not adequately compensate for the unsatisfactory distributional properties of D'. In conclusion, the test statistic D' may not produce reliable results for the analysis of the European data, and therefore, the following standardized test statistics are solely explored for the African expansion model.

3.2.2 FU AND LI'S D

FU and LI (1993) introduced several test statistics that unify the unbiased estimator of θ, based on the number of singletons, ξ_1, with the unbiased estimators of θ, based on S_n and Π_n, respectively. The authors also established test statistics with respect to the earlier introduced η_1, in case an outgroup is unavailable. Here, we exemplarily bring the statistical test that relates ξ_1 to S_n into the non-equilibrium background. First, we set up the necessary equations, which follow from Equations 2.23, 2.24 and 2.25, respectively. Let

$$u_{n,k} = \frac{k(k-1)}{n-1} \quad \text{and} \quad v_{n,k} = \frac{(n-k)(n-k-1)}{(n-1)(n-2)}.$$

Then,

$$E(\xi_1) = \frac{\theta}{2} \sum_{k=2}^{n} u_{n,k} E(T_k), \tag{3.6}$$

$$V(\xi_1) = \begin{cases} \frac{\theta}{2} \sum_{k=2}^{n} u_{n,k} E(T_k) + \frac{\theta^2}{4} \sum_{k=2}^{n} u_{n,k}(u_{n,k} V(T_k) + v_{n,k} E(T_k^2)) + \\ \frac{\theta^2}{2} \sum_{k=2}^{n-1} \sum_{k'=k+1}^{n} u_{n,k}(u_{n,k'} \text{Cov}(T_{k'}, T_k) + v_{n,k'} E(T_{k'} T_k)), & n > 2, \\ \theta E(T_2) + \theta^2 V(T_2), & n = 2, \end{cases} \tag{3.7}$$

$$\text{Cov}(\xi_1, S_n) = \frac{\theta}{2} \sum_{k=2}^{n} u_{n,k} E(T_k) + \frac{\theta^2}{4} \Big(\sum_{k=2}^{n} k u_{n,k} V(T_k)$$

$$+ \sum_{k=2}^{n-1} \sum_{k'=k+1}^{n} u_{n,k}(u_{k,k'} + k') \text{Cov}(T_{k'} T_k) \Big). \tag{3.8}$$

3.2. Generalization of classical test statistics

The equality

$$\theta = \frac{2E(S_n)}{E(T_c)} = \frac{2E(\xi_1)}{\sum_{k=2}^{n} u_{n,k} E(T_k)}$$

provides the summary statistic

$$d^* = f_{\xi_1} S_n - f_{S_n} \xi_1, \tag{3.9}$$

where $f_{\xi_1} = \sum_{k=2}^{n} u_{n,k} E(T_k)/2$, $f_{S_n} = E(T_c)/2$. The mean of d^* is 0 and the variance of d^* is given by

$$V(d^*) = f_{\xi_1}^2 V(S_n) + f_{S_n}^2 V(\xi_1) - 2 f_{\xi_1} f_{S_n} \mathrm{Cov}(\xi_1, S_n). \tag{3.10}$$

By making use of Equations 2.12, 3.7 and 3.8 together with the unbiased estimators of θ and θ^2, as given by Equations 3.3 and 3.4, one obtains the test statistic

$$D^* = \frac{d^*}{\sqrt{\hat{V}(d^*)}}, \tag{3.11}$$

where $\hat{V}(d^*)$ is the estimator of $V(d^*)$ with respect to Equations 3.3 and 3.4.
For the African demography ($n = 12$), we obtain

$$D^* = \frac{0.341 S_{12} - 0.758 \xi_1}{\sqrt{(0.1297 + 0.0123 S_{12}) S_{12}}}.$$

In complete analogy to the analysis of D', the distribution of D^* is generated for different values of θ by coalescent simulations, non-polymorphic iterations are disregarded, and D^* is recalculated and re-evaluated for all the different sample sizes that occur in the African data. The distributions of D^* for different values of θ and for $n = 12$ and $n = 40$ are shown in Figure 3.5. The average D^* that incorporates the effect of population expansion is -0.15 in the African dataset, whereas Fu and Li's D results in an average of -0.77. The usage of D^* without considering the expansion in population size results in a false-positive rate of about 15%. All loci, which are significant with respect to the outermost 5%- and 95%-quantiles (underlined in Table 3.4) are shown in Table 3.5. As before, the p−values are obtained for various values of θ. When comparing loci, which are significantly negative with respect to D' and D^* (indicated by an asterisk in Table 3.5), we cannot assign a preference to any of these test statistics. However, as already described in Chapter 1, singletons have to be considered with caution, since they can result from different types of directional selection. Therefore, a generalized version of Tajima's D that disregards singletons is analyzed below.

Chapter 3. Testing neutrality under variable population size

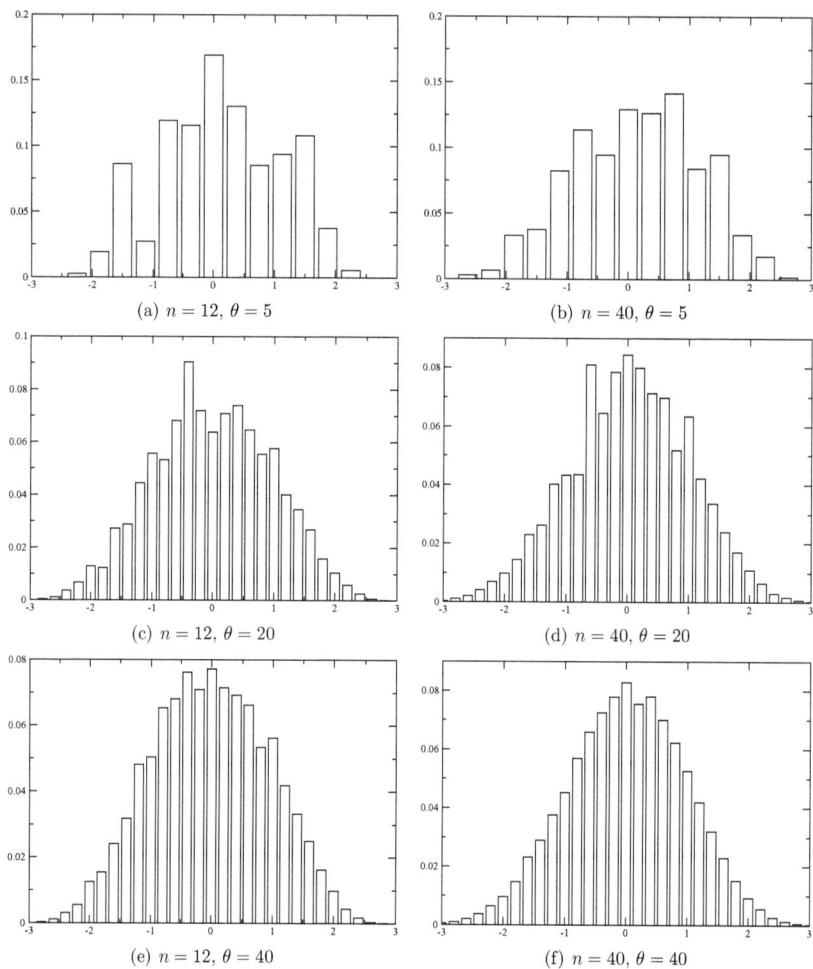

FIGURE 3.5 Simulated distributions of D^* for the African scenario.

3.2. Generalization of classical test statistics

TABLE 3.4 Distributional properties of D^* for the African scenario and $n = 12$.

θ	Mean	Variance	*0.05	*0.95
1	−0.02	1.00	−1.50	1.23
5	−0.05	0.99	−1.76	1.45
6	−0.05	0.99	<u>−1.77</u>	1.46
8	−0.06	0.98	−1.76	<u>1.57</u>
10	−0.07	0.97	−1.76	1.57
20	−0.09	0.95	−1.70	1.50
40	−0.12	0.92	−1.69	1.46
60	−0.13	0.91	−1.69	1.46

Results are obtained from coalescent simulations with 100,000 replicates.
* 5%- and 95%- quantile, respectively.

TABLE 3.5 List of outlier loci with respect to D^* in the African sample.

locus	160	163	419	728	323	*743	*430	*470	*122	729	10
D^*	−1.80	−1.81	−1.97	−1.97	−2.01	−1.95	−2.18	−2.22	−2.26	1.67	2.14
p−value: $\theta = 6$	0.028	0.028	0.012	0.012	0.012	0.010	0.008	0.002	0.002	0.026	0.002
$\theta = 10$	0.034	0.034	0.019	0.019	0.019	0.020	0.008	0.004	0.004	0.034	0.003
$\theta = 20$	0.037	0.037	0.024	0.024	0.021	0.024	0.012	0.008	0.008	0.031	0.003
$\theta = 40$	0.039	0.036	0.023	0.023	0.019	0.021	0.011	0.009	0.008	0.027	0.003
n	12	12	12	12	12	11	12	11	12	10	12

Results are obtained from coalescent simulations with 100,000 replicates.

Chapter 3. Testing neutrality under variable population size

3.2.3 FAY AND WU'S H

An excess of high-frequency derived variants is a unique consequence of positive directional selection (FAY and WU 2000). As already explained above on the basis of the theoretical result for the frequency spectrum, this excess cannot emerge due to variations in population size. It can be neither explained by background selection, whose two main consequences are the reduction of the effective size of the population and a skew in the site-frequency spectrum towards low-frequency derived alleles (e.g., WAKELEY 2008). FAY and WU (2000) introduced an estimator of θ, defined as

$$\hat{\theta}_H = \frac{1}{\binom{n}{2}} \sum_{i=1}^{n-1} i^2 \xi_i.$$

This estimator is constructed to weight high-frequency derived variants through the multiplication by i^2 and to be unbiased under the standard neutral model. Since this statistical quantity is not unbiased under general conditions, it has to be rescaled by

$$f_H = \sum_{k=2}^{n} \frac{2n-k+1}{(n-1)(k+1)} E(T_k), \qquad (3.12)$$

such that $\hat{\theta}_H/f_H$ becomes an unbiased estimator of θ. Note that f_H equals one under constant population size. First, we deduce that the rescaled version is an unbiased estimator of θ for general coalescent trees. It is simple to show that

$$\sum_{i=1}^{n-1} i^2 \frac{\binom{n-i-1}{k-2}}{\binom{n-1}{k-1}} = \frac{n(2n-k+1)}{(k+1)k}. \qquad (3.13)$$

Applying Equations 2.13 and 2.23, we obtain

$$E(\hat{\theta}_H)/f_H = \frac{1}{\binom{n}{2}} \sum_{i=1}^{n-1} i^2 E(\xi_i)/f_H = \frac{\theta}{2} \sum_{k=2}^{n} \frac{1}{\binom{n}{2}} \Big(\sum_{i=1}^{n-1} i^2 \frac{\binom{n-i-1}{k-2}}{\binom{n-1}{k-1}} \Big) k E(T_k)/f_H.$$

Applying Equation 3.13, it follows that $E(\hat{\theta}_H)/f_H = \theta$. The variance of $\hat{\theta}_H$ can be evaluated based on

$$\begin{aligned}
V(\hat{\theta}_H) &= \frac{1}{\binom{n}{2}^2} V\Big(\sum_{i=1}^{n-1} i^2 \xi_i\Big) \\
&= \frac{1}{\binom{n}{2}^2} \Big(\sum_{i=1}^{n-1} i^4 V(\xi_i) + 2 \sum_{i=1}^{n-2} \sum_{j=i+1}^{n-1} i^2 j^2 \mathrm{Cov}(\xi_i, \xi_j) \Big)
\end{aligned}$$

and the results of Chapter 2 for $V(\xi_i)$ and $\mathrm{Cov}(\xi_i, \xi_j)$.

3.2. Generalization of classical test statistics

As we have already seen above, $E(\Pi_n) = \theta f_{\Pi_n}$, where $f_{\Pi_n} = E(T_2)_2$ under variable population size. Joining the two different estimators of θ, based on Π_n and $\hat{\theta}_H$, we obtain

$$H' = f_H \Pi_n - f_{\Pi_n} \hat{\theta}_H. \tag{3.14}$$

Note that, although H' simplifies to H under constant population size, we use a slightly different definition of H' compared to the originally proposed version by FAY and WU (2000), where we would have to subtract $\hat{\theta}_H/f_H$ from Π_n/f_{Π_n}. Since the ranking of the simulated values is exactly the same under both definitions of H', the corresponding p–values for the resulting quantiles are independent from the chosen definition. For the African scenario ($n = 12$), we obtain

$$H' = 0.203 \Pi_{12} - 0.228 \hat{\theta}_H,$$

whereas for the European scenario ($n = 12$), the test statistic is given by

$$H' = 1.152 \Pi_{12} - 0.807 \hat{\theta}_H.$$

Simulated distributions for various values of θ and both demographic scenarios are illustrated in Figure 3.6. For both scenarios, and in particular for the European one, we find outliers in the data, but only if we estimate $\hat{\theta}$ based on S_n for each locus. There is just one locus (bold in Table 3.6) that appears as an outlier for the whole range of estimated θ-values (approx. $0.3 - 10$ for the European sample). It might appear adequate to estimate θ individually for each locus, but for genomic scans it is desirable to construct statistical tests that are more insensitive to the choice of θ, unlike here for H'.

TABLE 3.6 List of outlier loci with respect to H' in both samples.

	Africa					Europe											
locus	**483**	392	276	310	295	346	375	228	221	196	225	319	276	186	903	384	251
H'	−3.70	−0.83	−1.08	−0.93	−1.41	−4.61	−1.81	−1.29	−5.06	−1.29	−5.90	−2.16	−2.16	−8.29	−8.35	−7.54	−6.44
*$\hat{\theta}$	58.08	9.24	11.88	9.24	10.56	3.12	0.89	0.45	3.12	0.45	3.57	0.89	0.89	4.91	3.47	3.12	2.23
**p–value	0.035	0.032	0.027	0.026	0.011	0.045	0.039	0.038	0.038	0.038	0.036	0.032	0.032	0.032	0.019	0.017	0.012
n	12	12	12	12	12	12	12	12	12	12	12	12	12	12	21	12	12

Results are obtained from coalescent simulations with 100,000 replicates.
*$\hat{\theta}$ estimated per locus based on S_n.
**p–value refers to the per-locus value of $\hat{\theta}$.

Chapter 3. Testing neutrality under variable population size

Nevertheless, it is surprising that few loci appear as outliers, even if one takes the locally estimated θ into account, since the observed and estimated frequency spectra show a distinct difference in terms of high-frequency derived alleles (cf. Figure 3.2). This issue is discussed in the following section.

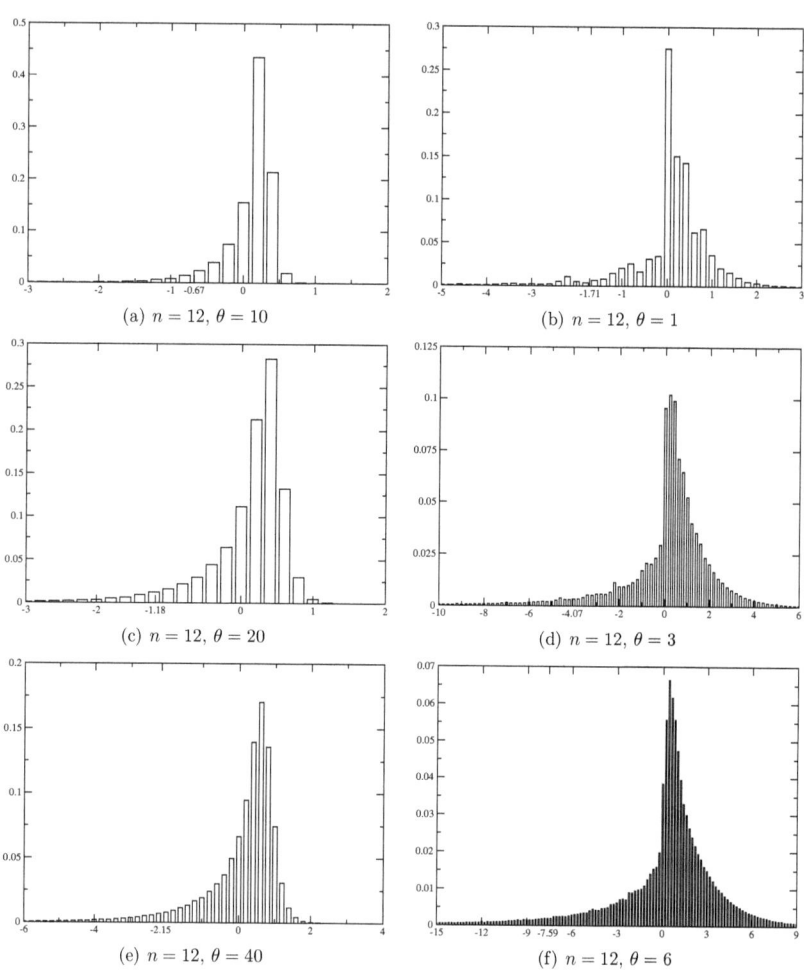

FIGURE 3.6 Simulated distributions of H' for the African (left panels) and European (right panels) scenario. Non-integer numbers show the 5%-quantiles.

3.2. Generalization of classical test statistics

3.2.4 THE SINGLETON-EXCLUSIVE VERSION OF TAJIMA'S D

Recently, ACHAZ (2008) proposed two modified versions of Tajima's D for the standard neutral model that exclude singletons or mutations of type 1. Here, we formulate the singleton-exclusive version of Tajima's D for general coalescent trees. Let $S_n^{-\xi_1}$ and $\Pi_n^{-\xi_1}$ be the number of segregating sites and the average number of pairwise differences in a sample of size $n \geq 3$, respectively, where singletons are disregarded. Using Equations 2.11, 2.26 and 3.6, it is simple to obtain

$$E(S_n^{-\xi_1}) = E(S_n) - E(\xi_1) = \frac{\theta}{2} \sum_{k=2}^{n} \frac{k(n-k)}{n-1} E(T_k), \tag{3.15}$$

$$E(\Pi_n^{-\xi_1}) = E(\Pi_n) - \frac{2}{n} E(\xi_1) = \frac{\theta}{2} \sum_{k=2}^{n} \frac{2(n-k)(n+k+1)(k-1)}{n(n-1)(k+1)} E(T_k). \tag{3.16}$$

The corresponding second-order moments can be written as

$$V(S_n^{-\xi_1}) = V(\sum_{i=2}^{n-1} \xi_i) = \sum_{i=2}^{n-1} V(\xi_i) + 2\sum_{i=2}^{n-2}\sum_{j=i+1}^{n-1} \mathrm{Cov}(\xi_i, \xi_j),$$

$$V(\Pi_n^{-\xi_1}) = V(\frac{1}{\binom{n}{2}} \sum_{i=2}^{n-1} i(n-i)\xi_1)$$

$$= \frac{1}{\binom{n}{2}^2} \Big(\sum_{i=2}^{n-1} i^2(n-i)^2 V(\xi_i) + 2\sum_{i=2}^{n-2}\sum_{j=i+1}^{n-1} i(n-i)j(n-j)\mathrm{Cov}(\xi_i, \xi_j)\Big),$$

$$\mathrm{Cov}(S_n^{-\xi_1}, \Pi_n^{-\xi_1}) = \frac{1}{\binom{n}{2}} \Big(\sum_{i=2}^{n-1} i(n-i)V(\xi_i) + \sum_{i=2}^{n-2}\sum_{j=i+1}^{n-1} (i(n-i)+j(n-j))\mathrm{Cov}(\xi_i, \xi_j)\Big).$$

Applying Equations 2.24 and 2.25, these equations can be written in terms of the first- and second-order moments of waiting times. We only give the solution for $V(S_n^{-\xi_1})$ in terms of moments of T_k, since $V(\Pi_n^{-\xi_1})$ and $\mathrm{Cov}(S_n^{-\xi_1}, \Pi_n^{-\xi_1})$ cannot be as elegantly represented as $V(\Pi_n)$ and $\mathrm{Cov}(S_n, \Pi_n)$. Let

$$g_{n,k} = \frac{k(n-k)}{n-1} \quad \text{and} \quad h_{n,k} = \frac{(k-1)(n-k-1)}{(n-1)(n-2)}.$$

Then,

$$V(S_n^{-\xi_1}) = \frac{\theta}{2} \sum_{k=2}^{n} g_{n,k} E(T_k) + \frac{\theta^2}{4} \sum_{k=2}^{n} g_{n,k}(g_{n,k}V(T_k) + h_{n,k}E(T_k^2)) + \tag{3.17}$$

$$\frac{\theta^2}{2} \sum_{k=2}^{n-1} \sum_{k'=k+1}^{n} g_{n,k'}(g_{n,k}\mathrm{Cov}(T_{k'}, T_k) + \frac{k(k-1)}{k'(k'-1)} h_{n,k'} E(T_{k'}T_k)).$$

Chapter 3. Testing neutrality under variable population size

The equality

$$\theta = \frac{E(\Pi_n^{-\xi_1})}{\sum_{k=2}^{n} \frac{(n-k)(n+k+1)(k-1)}{n(n-1)(k+1)} E(T_k)} = \frac{2E(S_n^{-\xi_1})}{\sum_{k=2}^{n} g_{n,k} E(T_k)}$$

suggests the statistic

$$d'_{-\xi_i} = f_{S_n^*} \Pi_n^{-\xi_1} - f_{\Pi_n^*} S_n^{-\xi_1}, \qquad (3.18)$$

where $f_{S_n^*} = \sum_{k=2}^{n} g_{n,k} E(T_k)/2$, $f_{\Pi_n^*} = \sum_{k=2}^{n} \frac{(n-k)(n+k+1)(k-1)}{n(n-1)(k+1)} E(T_k)$. The mean of $d'_{-\xi_i}$ is 0 and the variance of $d'_{-\xi_i}$ is given by

$$V(d'_{-\xi_i}) = f_{S_n^*}^2 V(\Pi_n^{-\xi_1}) + f_{\Pi_n^*}^2 V(S_n^{-\xi_1}) - 2 f_{S_n^*} f_{\Pi_n^*} \mathrm{Cov}(S_n^{-\xi_1}, \Pi_n^{-\xi_1}). \qquad (3.19)$$

The singleton-exclusive, unbiased estimators of θ and θ^2, based on $S_n^{-\xi_1}$, can be derived via Equations 3.15 and 3.17. They are

$$\hat{\theta}_{-\xi_1} = \frac{2 S_n^{-\xi_1}}{\sum_{k=2}^{n} g_{n,k} E(T_k)}, \qquad (3.20)$$

$$\hat{\theta^2}_{-\xi_1} = \frac{4 S_n^{-\xi_1}(S_n^{-\xi_1} - 1)}{\sum_{k=2}^{n} g_{n,k}(g_{n,k} + h_{n,k}) E(T_k^2) + 2 \sum_{k=2}^{n-1} \sum_{k'=k+1}^{n} g_{n,k'}(g_{n,k} + \frac{k(k-1)}{k'(k'-1)} h_{n,k'}) E(T_{k'} T_k))}. \qquad (3.21)$$

The simplification of Equation 3.19 in terms of moments of T_k, and replacing θ and θ^2 by Equations 3.20 and 3.21 in the resulting formula, leads to $\hat{V}(d'_{-\xi_i})$. Hence, we obtain

$$D'_{-\xi_1} = \frac{d'_{-\xi_1}}{\sqrt{\hat{V}(d'_{-\xi_1})}}. \qquad (3.22)$$

Before we apply the test statistic $D'_{-\xi_1}$ to the African demography, we first note $V(\Pi_n^{-\xi_1})$ and $\mathrm{Cov}(S_n^{-\xi_1}, \Pi_n^{-\xi_1})$ for the standard neutral model. Let

$$h_1 = \sum_{i=1}^{n-1} \frac{1}{i} \quad \text{and} \quad h_2 = \sum_{i=1}^{n-1} \frac{1}{i^2}.$$

Then,

$$V(\Pi_n^{-\xi_1}) = \theta\left(\frac{n+1}{3(n-1)} - \frac{4}{n^2}\right) +$$
$$\theta^2 \frac{2(n(n(n(n-1) - 107) + 246) + 72n(n-2)h_2 - 36(n-4)h_1 - 144)}{9n^2(n-1)(n-2)},$$

3.2. Generalization of classical test statistics

and

$$\mathrm{Cov}(S_n^{-\xi_1}, \Pi_n^{-\xi_1}) = \theta \frac{n-2}{n} + \theta^2 \frac{(n-1)(n(n-12)+4) + 8n(n-2)h_2 - 4(n-6)h_1}{2n(n-1)(n-2)}.$$

These two equations contradict the corresponding formulas in ACHAZ (2008). The reason is that the derivation of the singleton- and type-1-exclusive versions of Tajima's D in ACHAZ (2008) relied on the erroneous equation for $\mathrm{Cov}(\xi_1, \Pi_n)$ in FU and LI (1993). Curiously, the correct version of $\mathrm{Cov}(\xi_1, \Pi_n)$ can be derived on the basis of Equation 12 in FU (1995), and is given by

$$\mathrm{Cov}(\xi_1, \Pi_n) = \theta \frac{2}{n} + \theta^2 \left(\frac{6}{n} - \frac{4(nh_2 - h_1)}{n(n-1)} \right).$$

The elimination of singletons in the African dataset results in five non-polymorphic loci (cf. Table 3.7). It is probably not surprising that many of these loci deviate significantly ($p < 0.05$) from the expansion model with respect to the test statistics D' and particularly D^*.

To apply $D'_{-\xi_1}$ to the African demography, we adapt the demographic parameters of LI and STEPHAN (2006) that had been estimated by disregarding singletons. Then, the estimated time of the expansion is 0.033 (in units of $2\hat{N}_{A0}$) in the past, and the population size before the expansion is about 0.15 of the actual size. To compare the outcome of the theoretical result for these parameters with the African data, we require the frequency spectrum without singletons. The probability $q^*_{n,i}$ of the number of times, i, a single mutation arising between the present and the time of the most recent common ancestor, is represented in the sample of size $n > 3$, as $\mu \to 0$, is given by

$$q^*_{n,i} = \frac{(n-i-1)!(i-1)! \sum_{k=2}^{n-i+1} k(k-1)\binom{n-k}{i-1} E(T_k)}{(n-2)! \sum_{k=2}^{n} k(n-k) E(T_k)}, \quad 1 < i < n. \tag{3.23}$$

This result is obtained by a trivial modification in the elegant derivation of Equation 2.1 by GRIFFITHS and TAVARÉ (2003). Although the theoretical frequency spectrum for the above parameters appears to not fit the African sample (cf. Figure 3.7) as well as before (cf. Figure 3.2), the method of LI and STEPHAN (2006) provides a reliable parameter estimate—at least for the case that only a single population size change is assumed. However, the instantaneous expansion model matched the observation significantly better than the standard neutral model (LI and STEPHAN 2006), although with less power. The decrease in power is due to the smaller amount of polymorphisms in the African dataset when singletons are disregarded. With the suggested African demography the analysis of $D'_{-\xi_1}$ is accomplished in complete analogy to the above-mentioned statistical tests.

Chapter 3. Testing neutrality under variable population size

TABLE 3.7 List of non-polymorphic loci when singletons are disregarded.

locus	122	470	419	728	745
D'	✓	✓			
D^*	✓	✓	✓	✓	

✓: $p < 0.05$

For the estimated parameters and $n = 12$, we obtain

$$D'_{-\xi_1} = \frac{0.323\Pi_{12}^{-\xi_1} - 0.132 S_{12}^{-\xi_1}}{\sqrt{(0.00071 + 0.00044 S_{12}^{-\xi_1}) S_{12}^{-\xi_1}}}. \qquad (3.24)$$

Simulated distributions of $D'_{-\xi_1}$ for different values of θ and, as before, for $n = 12$ and $n = 40$ are illustrated in Figure 3.8. The $D'_{-\xi_1}$-values of the African loci are overall very similar, regardless of the choice of the expansion or the standard neutral model. The average $D'_{-\xi_1}$ among loci is -0.33 for the expansion model and -0.36 for the standard neutral model. Although the standard neutral model has been rejected in favor of the expansion model by LI and STEPHAN (2006), we may conclude that adapting the standard neutral model into the test statistic $D'_{-\xi_1}$ results only in a minor difference. This issue will be discussed below.

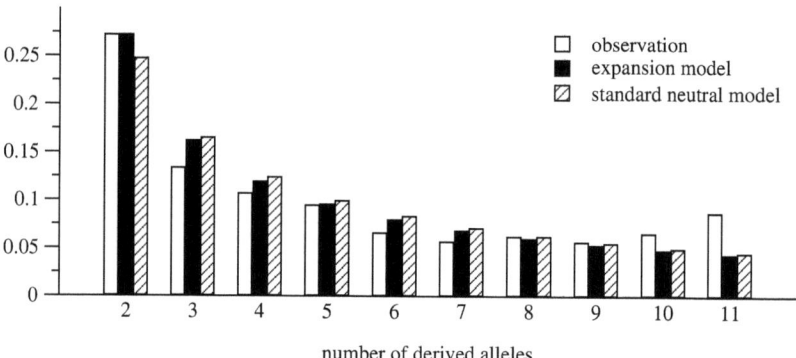

FIGURE 3.7 Comparison of the African data with the theoretical frequency spectra of the expansion and the standard neutral model.

3.2. Generalization of classical test statistics

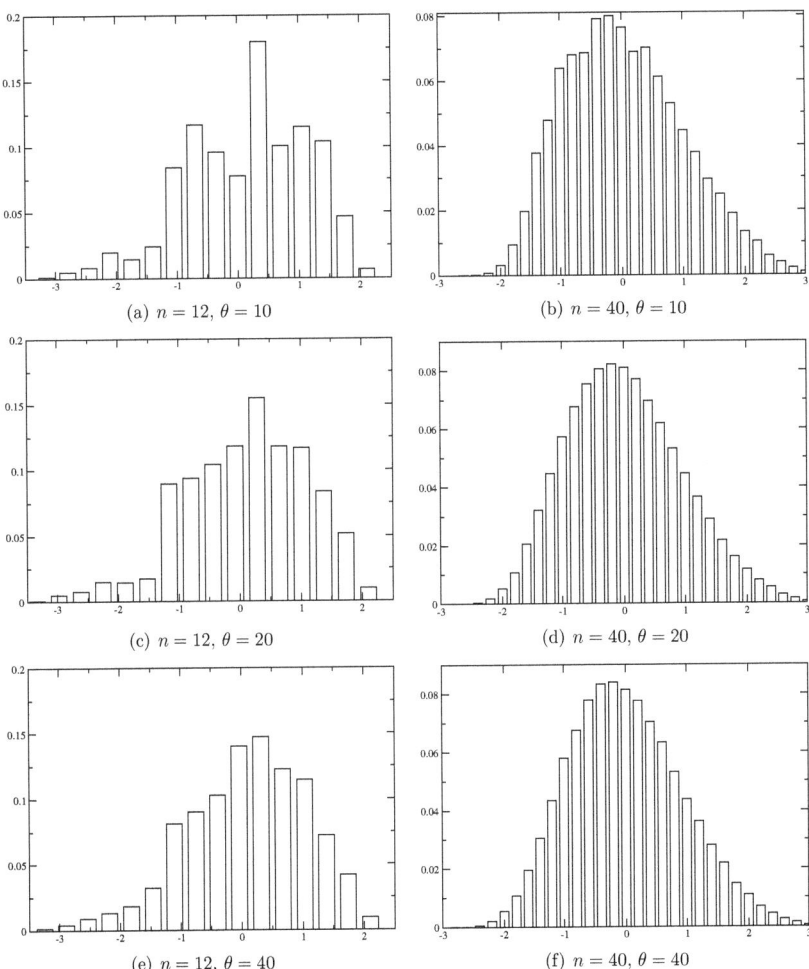

FIGURE 3.8 Simulated distributions of $D'_{-\xi_1}$ for the African scenario.

Chapter 3. Testing neutrality under variable population size

TABLE 3.8 Distributional properties of $D'_{-\xi_1}$ for the African scenario and $n = 12$.

θ	Mean	Variance	*0.05	*0.95
3	−0.027	1.02	−1.42	1.39
5	−0.041	1.02	−1.47	1.39
10	−0.052	1.01	−1.65	<u>1.45</u>
20	−0.078	0.98	−1.71	1.43
30	−0.086	0.96	<u>−1.74</u>	1.42
50	−0.094	0.93	−1.74	1.38
70	−0.101	0.92	−1.74	1.36
100	−0.103	0.91	−1.74	1.35

Results are obtained from coalescent simulations with 100,000 replicates.
* 5%- and 95%- quantile, respectively.

TABLE 3.9 List of outlier loci with respect to $D'_{-\xi_1}$ in the African sample.

locus	$D'_{-\xi_1}$	p−value				*D'	n
		$\theta = 10$	$\theta = 20$	$\theta = 40$	$\theta = 80$		
329	−2.40	0.044	0.038	0.031	0.027	−1.30	10
743	−1.97	0.050	0.048	0.045	0.045	−1.64	11
447	−1.82	0.049	0.047	0.045	0.044	−1.43	12
25	−1.88	0.048	0.044	0.041	0.041	−1.54	12
392	−1.88	0.048	0.044	0.041	0.041	−1.15	12
26	−1.90	0.045	0.042	0.040	0.040	−1.40	12
276	−1.91	0.045	0.041	0.040	0.040	−0.97	12
430	−1.99	0.043	0.038	0.036	0.035	−1.74	12
725	−1.99	0.043	0.038	0.036	0.035	−1.06	12
310	−2.44	0.014	0.017	0.017	0.016	−1.40	12
295	−3.15	0.0017	0.0025	0.0023	0.0016	−1.70	12
451	1.51	0.043	0.039	0.034	0.030	0.38	12
472	1.53	0.036	0.035	0.032	0.027	−0.35	12

Results are obtained from coalescent simulations with 100,000 replicates.
*D' regarding the singleton-exclusive demographic estimates.

3.2. Generalization of classical test statistics

As before for D' and D^*, non-polymorphic iterations are disregarded, and the outermost 5%- and 95%-quantiles among simulation results based on different values of θ (cf. Table 3.8) are used for the selection of outliers (cf. Table 3.9). The range of possible θ-values has been estimated by Equation 3.20 and the two outermost values are approximately 3 and 100, respectively. In contrast to D', we obtain an increased amount of candidate loci. This may be surprising at first sight, since one would rather expect a decrease of significant loci due to the exclusion of singletons. One might hypothesize that this is due to the slight change in the demographic estimates. However, when adapting these estimates to D' and the full dataset, there are only a few loci (bold in Table 3.9) that would be significant with respect to D'. The explanation for this outcome is fairly simple. As already pointed out by ACHAZ (2008), the singleton-exclusive version of Tajima's D outperforms the original version with respect to high-frequency derived alleles, and this holds alike for the generalized version. The problem of D' to detect an excess of high-frequency derived alleles in the presence of low-frequency derived polymorphisms is considerably reduced by the removal of singletons. As a simple example, imagine a sample of size 12 that contains two mutations, one of size 1 and one of size 11. Then, $D' = -1.16$ and $D'_{-\xi_1} = -2.30$ under the same demographic scenario that had been estimated by disregarding singletons (LI and STEPHAN 2006). As illustrated in Table 3.9, even a change of the algebraic sign is possible and exemplifies an alteration in the ranking of $D'_{-\xi_1}$-values, compared to D'-values (cf. Figure 3.9). Furthermore, Table 3.9 demonstrates that the p-values are fairly insensitive to different values of θ, in contrast to what was seen before for H'. A singleton-exclusive, preferably standardized version of H' is expected to perform better for the detection of high-frequency derived variants. However, this will be analyzed in more detail elsewhere.

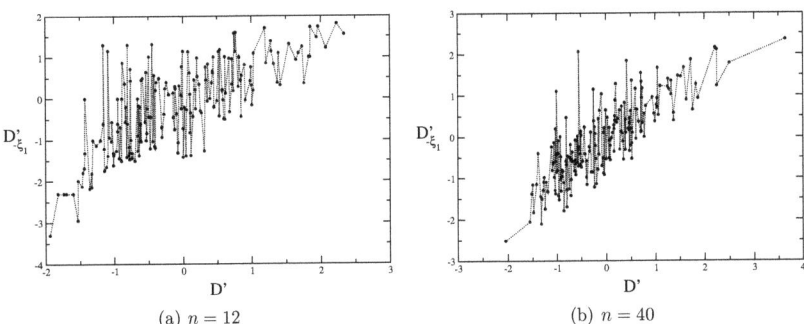

FIGURE 3.9 The two plots are obtained from coalescent simulations with 200 iterations under the estimated demographic scenario, when singletons are disregarded. x-axis: D'-values sorted by their rank. y-axis: $D'_{-\xi_1}$-values with respect to the outcome for D'.

Chapter 3. Testing neutrality under variable population size

TABLE 3.10

	expansion model				standard model		
$\hat{\theta}_{-\xi_1}$	$V(S_{12}^{-\xi_1})$	$E(\Pi_{12}^{-\xi_1})$	$V(\Pi_{12}^{-\xi_1})$	$\hat{\theta}_{-\xi_1}$	$V(S_{12}^{-\xi_1})$	$E(\Pi_{12}^{-\xi_1})$	$V(\Pi_{12}^{-\xi_1})$
	theoretical single-locus results						
	36.51	3.94	7.36		38.58	3.98	7.83
	experimental data						
29.87	43.62	3.72	6.96	4.78	43.62	3.72	6.96

* subset of 202 loci ($n = 12$, $\bar{L} = 475$), where the outgroup is available. Due to varying locus lengths, the experimental outcomes of S_{12} and Π_{12} are rescaled with respect to $\bar{L}$. $\hat{\theta}_{-\xi_1}$ is estimated on the subset of 202 loci for both models, and thereafter used for the theoretical results.

As already mentioned above, the difference between the expansion model and the standard neutral model is smaller when singletons are excluded than when considering the entire amount of polymorphisms in the African sample. This might not be surprising, since an excess of singletons is the most pronounced feature of an expanding population. However, both models explain the experimental data fairly well in terms of S_n and Π_n (cf. Table 3.10). In particular, there is a much smaller discrepancy in the variance of S_n, compared to the outcome for the entire set of polymorphism data (cf. Table 3.1). One may now be tempted to replace the expansion model by the standard neutral model, since the summary statistics for both models are in good agreement with the data; but see Figure 3.11. Furthermore, one might argue that the incorporation of purifying selection into the standard neutral model would lead to an increase in the amount of doubletons as well, such that the resulting nearly neutral model of constant population size may not be rejected in favor of the expansion model anymore. However, in terms of the detection of outliers based on $D'_{-\xi_1}$ it makes a marginal difference on which model we would rely on (cf. Figure 3.10). Since the estimated $\hat{\theta}_{-\xi_1}$ of the standard neutral model is reduced by a factor of approximately 6.26 (cf. Table 3.10), the $p-$values generated under this model remain virtually the same (cf. Table 3.11).

It is encouraging that with help of the test statistic $D'_{-\xi_1}$, one can retrieve an excess of high-frequency derived alleles, that cannot clearly be detected by D', since singletons conceal their signature.

3.2. Generalization of classical test statistics

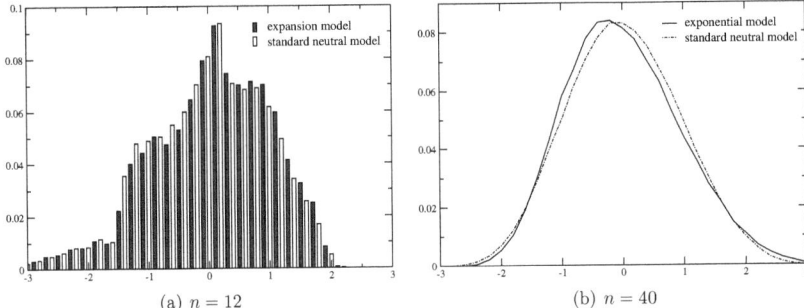

FIGURE 3.10   Comparison of the simulated distributions of $D'_{-\xi_1}$ for the expansion model, where $\theta = 40$ for both sample sizes, and the standard neutral model, where $\theta = 6.4$ and $\theta = 7.3$ for $n = 12$ and $n = 40$, respectively. The differences in the θ-values for both scenarios refer to the differences in total tree length.

TABLE 3.11   Examples of outlier loci with respect to $D_{-\xi_1}$ for the standard neutral model.

locus	$D_{-\xi_1}$	p−value				n
		$\theta = 1.6$	$\theta = 3.2$	$\theta = 6.4$	$\theta = 12.8$	
430	−2.02	0.043	0.039	0.037	0.036	12
725	−2.02	0.043	0.039	0.037	0.036	12
310	−2.46	0.017	0.020	0.019	0.017	12
295	−3.16	0.0023	0.0030	0.0028	0.0019	12

Results are obtained from coalescent simulations with 100,000 replicates.

Chapter 3. Testing neutrality under variable population size

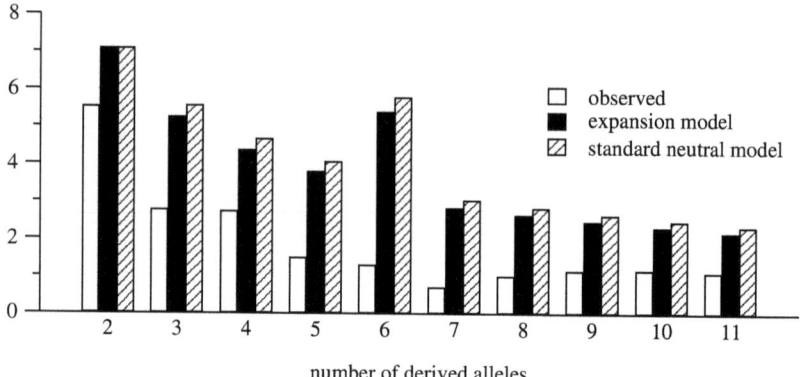

number of derived alleles

FIGURE 3.11 Comparison of both models with the 202 loci of the African data with $n = 12$ and available outgroup. The estimates $\hat{\theta}_{-\xi_1}$ (cf. Table 3.10) are used for the plots of the theoretical results under both models. The variance of ξ_i decreases with i, except for $i = n/2$ when n is even, as here.

We finally want to emphasize some caveats. It is important to not over-interpret the good fit of the theoretical results for S_n and Π_n of both models to the genome-wide assembled data (cf. Table 3.10). Both, S_n and Π_n, summarize the absolute site-frequencies ξ_i. In terms of the means of ξ_i, the expansion and the standard neutral model fit as well to the data (for $2 \leq i \leq 11$) as the frequency spectrum. In contrast, the variances of ξ_i for both models show a strong deviation from the observed data (cf. Figure 3.11). Since the theoretical outcomes of $V(\Pi_n)$ and $V(S_n)$ are in good agreement with the observation, and the theoretical results for $V(\xi_i)$ overestimate the observed data, $\text{Cov}(\xi_i, \xi_j)$ in the African data is larger than in the according theoretical results. Besides disregarding intra-locus recombination in the theoretical framework, this might simply reflect that there is a relatively small distance between a certain amount of loci in the African data, such that their polymorphism patterns are more strongly correlated than when these loci could be seen as independent. In conclusion, it is hard to assess to which degree the outcome in Figure 3.11 reflects issues in the quality of the demographic estimates for the real demographic history of the African sample.

Chapter 5

DISCUSSION

Changes in population size during evolutionary history can obscure the traces left by natural selection on DNA polymorphism. Methods to identify potential target sites of selective substitutions in a genome, which rely on the spatial distribution of polymorphic sites, may be severely misled if demographic properties are ignored. In particular, the identification of adaptive substitutions can be difficult because colonization of a new habitat or environmental changes and adaptation to them often occur simultaneously. It is therefore essential that the possible joint effect of demographic and adaptive factors is adequately considered in models and statistical methods for data analysis.

JENSEN et al. (2005) demonstrated that the composite-likelihood-ratio (CLR) test proposed by KIM and STEPHAN (2002), which compares the alternative hypotheses of genetic hitchhiking and neutral evolution under constant population size, loses power and suffers from false-positives for certain demographic parameters. It is of interest to compare the performance of the CLR test and the variance-to-mean ratio for different parameter constellations of a bottleneck model (cf. Figure 4.1). Obviously, the variance-to-mean ratio of total tree length, $V(T_c)/E(T_c)$, a measure for the dispersion of the probability distribution of T_c, serves as an indicator of "critical" bottleneck scenarios, i.e. those which lead to a high false-positive rate in the detection of selective sweeps, if no means for correction of demographic biases are taken. However, we have to be careful with the interpretation of this result with regard to two properties. First, the variance-to-mean-ratio has to be considered with respect to the standard neutral model, where e.g., $V(T_c) \approx E(T_c)$ for $n = 15$ (cf. Figure 4.1). For fixed parameters of a given demographic model, the variance-to-mean ratio decreases with increasing n, such that a numerical value of $V(T_c)/E(T_c)$ becomes less meaningful without the reference value of the standard neutral model. Second, the variance-to-mean ratio is directly applicable as a detection criterion for delicate demographic histories, if and only if the population size at time of sampling was not exceeded over certain time spans in the past, as for instance in the

67

Chapter 4. Discussion

estimated European demography. Considering a simple model of population decline, where the population size is always constant but experienced an instantaneous drop to its present size at some point in the past, always induces an increase in $V(T_c)$ and $V(T_c)/E(T_c)$ compared to the standard neutral model. In contrast, any expansion model leads to a reduction in $V(T_c)$ and $V(T_c)/E(T_c)$ compared to the model of constant population size. For the African demography, we obtain about a six-fold reduction of $V(T_c)/E(T_c)$ compared to the standard neutral model. A population bottleneck, which is a mixture of these two scenarios, causes a reduction in $V(T_c)$ with respect to the standard neutral model, as long the population size at time of sampling was never exceeded in the past. The variance-to-mean ratio of T_c can be greater than or less than for the model of constant population size and it reflects the severity of the bottleneck. Before we explain the relationship between $V(T_c)/E(T_c)$ and the proportion of false-positives in the CLR test in more detail, we shall first describe how the variance-to-mean ratio is still applicable, when the population size in the past is larger than the present population size. Therefore, we consider two cases of the above decline model: (i) the population experienced a 50-fold decrease in size at time 0.1 in the past;

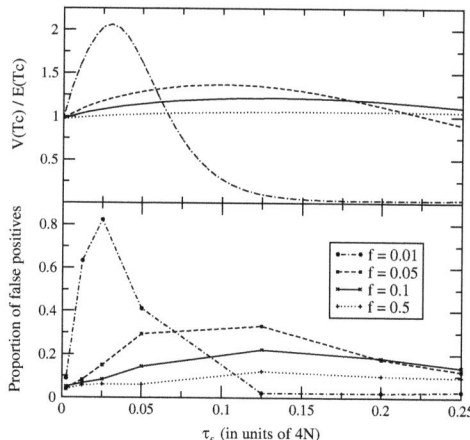

FIGURE 4.1 Comparison of the variance-to-mean ratio of total tree length and the proportion of false-positives in the CLR test (redrawn from Figure 1 in JENSEN et al. 2005) for various parameters of the following bottleneck model. At time τ_f in the past a population of previously constant size N is reduced to size $f \times N$ and then increases exponentially until reaching size N at time of sampling. The incorporation of exponential growth in this model requires a numerical evaluation of Equations 2.3 and 2.4 to obtain the theoretical graph for $V(T_c)/E(T_c)$. The sample size is 15 in both cases and $\theta = 75$ in the simulated graph of JENSEN et al. (2005). The scaled recombination rate $4Nr$ equals 0 and 1000 in the theoretical and simulated graph, respectively.

4 Discussion

(ii) the population experienced a 100-fold decrease in size at time 1.5 in the past. Let $n = 12$ in both cases and time be measured in units of two times the present population size. The results for $V(T_c)/E(T_c)$ are then about (i) 60 and (ii) 288. Now, we apply a simple time rescaling argument. Since we do not consider mutations at the moment, we can as well scale the time to decline and both population sizes with respect to the anterior population size instead of the present population size. Then the present population size is (i) 1/50 and (ii) 1/100 and the time to decline is (i) 0.002 and (ii) 0.015. To calculate $V(T_c)/E(T_c)$ for the rescaled versions, we consider the bottleneck model from Chapter 2 (cf. Figure 2.2) that consists of three phases and is defined as $\lambda(t) = f$ for $\tau \leq t < \tau + \tau_f$ and $\lambda(t) = 1$ otherwise. Note that throughout this chapter a population bottleneck always refers to this definition, except for the comparison of the proportion of false-positives in the CLR test and the variance-to-mean ratio of T_c as illustrated in Figure 4.1. For $\tau = 0$, we obtain (i) $V(T_c)/E(T_c) \approx 1.21$ and (ii) $V(T_c)/E(T_c) \approx 2.88$ by applying the corresponding formulas from Chapter 2. As we will see below, these results are completely in line with the feasibility to incorporate these two decline models into the test statistic D'. In conclusion, by scaling time with respect to the largest instead of the present population size, the variance-to-mean ratio of total tree length can also be used as an indicator of the statistical tractability of a population that experiences a decline in its size. To reconsider the relationship of $V(T_c)/E(T_c)$ and the false-positive-rate of the CLR-test, we just note that population bottlenecks with the largest values of $V(T_c)/E(T_c)$ introduce a skew in the frequency spectrum of segregating sites towards singletons and intermediate- to high-frequency derived variants, as well.

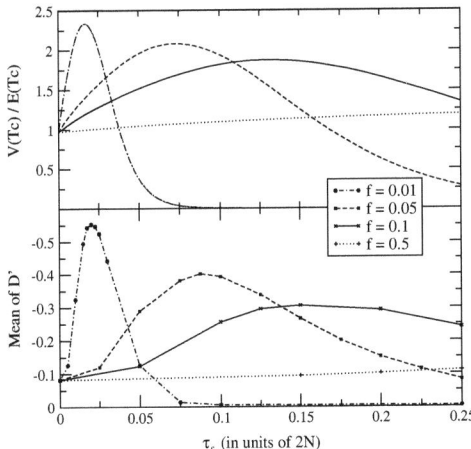

FIGURE 4.2 Variance-to-mean ratio of total tree lengths and means of D' for a population bottleneck. Fixed parameters are $n = 15$, $\tau = 0.02$ and $\theta = 10$.

Chapter 4. Discussion

These deviations from the standard neutral model are detected by the CLR-test, such that one obtains an appropriate proportion of false-positives, when genetic hitchhiking is considered as the alternative hypothesis. Taking recombination into account may cause a considerable reduction of $V(T_c)$, whereas $E(T_c)$ remains unaffected (HUDSON 1990). Therefore, the variance-to-mean ratio of T_c may be noticeably reduced, but still be elevated for critical bottlenecks with respect to the standard neutral model.

Despite our capability to rescale several standard neutral estimates of θ (cf. Chapter 3) for any given model of population size change, for certain demographic scenarios it is difficult, or even impossible, to filter out the effects on the statistics imposed by the demography. As we have seen in the analysis of the suggested European demography in Chapter 3, the distributional properties of the generalized test statistic D' appeared as too unsatisfactory for a reliable inference. To appreciate this effect quantitatively one may exemplarily compare $V(T_c)/E(T_c)$ with the mean of D' (cf. Figure 4.2). Not only can the mean of D' become heavily biased for problematic parameter constellations, but also the distribution of D' can become bimodal or even ragged (cf. Figure 4.3), which makes a meaningful definition of the rejection region difficult. Note that the inversion of the argument concerning the bias of $\overline{D'}$ does not hold. As illustrated in Figure 4.3, a demographic scenario that induces a small amount of observable polymorphism may conceal the shape of the genealogies with respect to the moments of D'.

The example of such a severe population bottleneck may be seen as appropriate to briefly comment upon the idea that traces of selective sweeps may be distinguished from those of recent population bottlenecks, since only the former have a local effect, while the latter should have a chromosome-wide effect (cf. Chapter 1 or similarly phrased in many population genetical publications). In the case of a severe population bottleneck, independent or loosely linked loci can be roughly separated into two classes. They can either find their MRCA during the phase of reduced population size or their ancestry may pass the point of decline such that their MRCA arises much further back in time. This means that loci of reduced variability and loci with a normal or even relatively high level of variability are scattered over the genome, such that, depending on their chromosomal

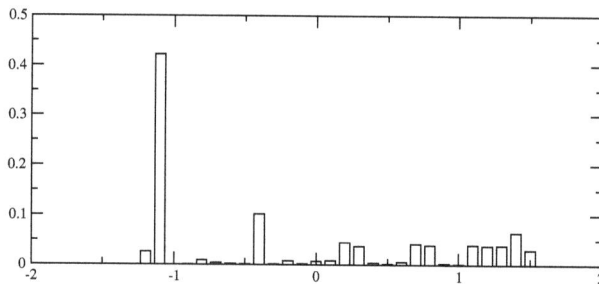

FIGURE 4.3 The distribution of D' for a population bottleneck. The parameters are $n = 15$, $\tau = 0.02$, $\tau_f = 0.002$, $f = 0.001$ and $\theta = 1$. Note that $\overline{D'} \approx -0.2$ and $V(D') \approx 1$.

4 Discussion

location, even v-shaped valleys of variability measures, as for instance Π_n, may appear and mimic the pattern that we would usually expect under a selective sweep. In conclusion, a severe population bottleneck has a genome-wide, but non-homogeneous, effect.

It is discouraging that bottleneck scenarios which already caused problems for tests of neutral evolution under constant population size, remain cumbersome when tested under a more general model of variable population size. However, the impact of varying population size on the observable pattern of genetic variability can be examined in many biologically relevant instances. If a population has experienced an expansion in its size, one identifies less outliers and reduces the false-positives compared to the commonly applied model of constant population size. In cases of population decline, and if previously estimated demographic parameters allow an accompanying application of the generalized test statistics introduced in Chapter 3, one obtains a larger number of candidate loci than if one would assume the standard neutral model. This is due to the fact that the standard version of Tajima's D tends to be negative for an expansion model, whereas the same test statistic becomes positive for population decline. Consequently, if Tajima's D is applied to data originating from an expanding population, too many outliers are found; e.g., $\lambda(t) = \exp\{-50t\}$ leads to a false-positive rate of over 60% for a sample of size 20. This exemplified scenario can be corrected for. In contrast, the application of Tajima's D to data from a population that experienced a decline will result in an underestimation of candidate loci.

Throughout Chapter 3 we have considered the demographic history of both *Drosophila* samples as given and subjected it to several generalized test statistics. In this context it is reasonable to raise the following questions. Can we tell from the shape of the frequency spectrum and without explicit knowledge of the demographic scenario, whether a certain statistical analysis is feasible or not? Is it sufficient to estimate a demographic scenario that provides a reasonable fit to the data or do different demographic estimates, that offer a similar goodness-of-fit, influence the test statistic D' in a different manner? The last question is posed with reference to the generalized version of Tajima's D, since its standard neutral version is the most commonly used of all the test statistics that have been generalized in Chapter 3. To address these questions, we distinguish three demographic models: (i) instantaneous population expansion; (ii) instantaneous population decline; and (iii) population bottleneck. On the basis of Figure 4.4, the characteristics of the frequency spectrum of an expansion and a recent decline model shall be considered. As already seen in Chapter 3, population expansion, as for instance in the African scenario, mainly results in an excess of singletons, whereas all the other site-frequencies are underrepresented compared to the standard neutral model. Moving the time of expansion towards the past may result in an overrepresentation of doubletons as well, but all the other site-frequencies remain beneath the standard neutral counterparts. However, this can be seen as an indication that a statistical analysis for a sample taken from a population that experienced an expansion in its size is in principle always warranted. It is worth mentioning that in cases of strong exponential growth, which lead to polymorphism patterns largely consisting of singletons and where nearly all values of Tajima's D are false-positives, increasing the sample size is instrumental for the applicability of D'. As for the second question, we can use the implementation of the theoretical results of Chapter 2 in *Mathematica* (WOLFRAM 1999), version 6.0, to immediately obtain another

Chapter 4. Discussion

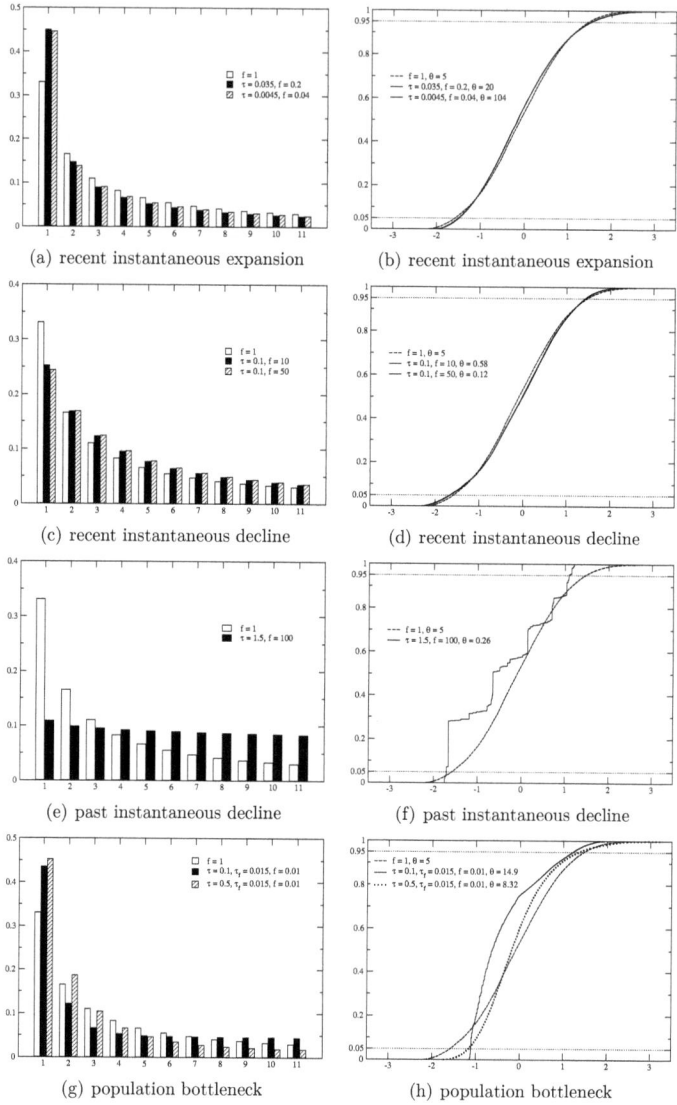

FIGURE 4.4 A sample of size 12 is traced backwards in time, N denotes the population size at time of sampling and $f = 1$ refers to the standard neutral model. The panels on the left constitute the frequency spectra, the panels on the right show simulated cumulative distributions of D'. (a)–(f): at point τ in the past the population size changes to $f \times N$. (g)–(h): at point τ in the past N reduces to $f \times N$ for the time span of τ_f, and reconverts instantaneously to N at time $\tau + \tau_f$. The various θ-values for the different demographic parameters refer to the estimates obtained by Equation 3.3, when an average of about 15 segregating sites is observed in the sample.

4 Discussion

instantaneous expansion model, as shown in (a), that fits well to the model proposed by LI and STEPHAN (2006). The parameters have been chosen to be distinctly different in terms of time of expansion and ancestral population size. Although the alternative parameters do not pretend to be historically correct, we encounter some difficulties in inferring the real demographic history of a population based on the frequency spectrum. Even if more population size changes are taken into account, various parameter constellations result in nearly indistinguishable frequency spectra. However, as demonstrated in (b), the distribution of D' for these two models provides essentially the same result. This suggests, on the one hand, that the original expansion model is robust to a different choice of parameters when the frequency spectra are similar enough. On the other hand, this example emphasizes that the frequency spectrum gives a more detailed picture than the D'-statistic, since the distributions of D' are fairly identical for these two expansion models, whereas the corresponding frequency spectra are more diverse. Both expansion models are similar in terms of first- and second-order moments of Π_n, S_n and the absolute site-frequencies, ξ_i, as well. This raises another question, which is closely related to the limited feasibility in obtaining the real population's demographic history based on the frequency spectrum.

How can one reliably infer the present population size? Both expansion models are similar with respect to all measures of DNA polymorphism and we could have chosen another parameter configuration, where the instantaneous expansion occurs further back in time from a relatively larger population size. As shown in Figure 4.4, both expansion models result in different estimates of θ, when we assume a sample with an average number of 15 segregating sites and apply Equation 3.3 to obtain $\hat{\theta}$. Since the average mutation rate per site per generation is the same for both models, we obtain about a five-fold difference for the present population size. To identify outliers based on the generalized test statistics of Chapter 3, it is irrelevant on which model we rely on. We solely emphasize that one should regard numerical estimates of present population sizes (cf. Chapter 3) with care as well as numerical estimates of a population's demographic history.

As pointed out in (c), recent population decline affects the frequency spectrum in the manner that a decay of singletons is observable, whereas the other site-frequencies are overrepresented compared to the standard neutral model. Furthermore, the strength of reduction has a rather small effect on the frequency spectrum and the distributions of D' are virtually the same for these two parameter constellations, as shown in (d). As long as the decline occurs recently, it is a "harmless" demographic event. The crucial difference to an expanding population is that the severity of this demographic model arises when we shift the time of decline into the past. As illustrated in (e), if an instantaneous decline from a distinctly larger population size occurs at point 1.5 in the past, the site-frequencies tend to be almost uniformly distributed and the distribution of D' becomes meaningless for defining a proper rejection region, as shown in (f). Population decline can be denoted as an awkward demographic scenario, when it occurs approximately between 0.5 and 4.5 in the past and if the population drops from a noticeably higher population size. The shape of the frequency spectrum can immediately tell that a statistical analysis aided by D' based on such a demographic scenario is not possible. This example of a severe population decline is of particular interest to reconsider the relativity of the terms recent and past, when time is scaled in units of two times the present population size.

Chapter 4. Discussion

Since the population does not recover from its decline in (e), the point of decline appears as relatively "old" compared to a bottleneck model that also includes the expansion to a large population size, as demonstrated in (g). Furthermore, the study of an instantaneous expansion and an instantaneous decline model reveals that population decline is the more "harmful" component, when both scenarios are merged into the bottleneck model, whereas the expansion to a higher population size rather conceals the profound effect of a bottleneck when considering the frequency spectrum. This effect is exemplified in (g) on the basis of a recent and a past bottleneck scenario. In both cases the duration of the bottleneck and the size of reduction equal the parameters in (e) in units of generations. Both parameter constellations lead to an excess of singletons. Consequently, for the recent bottleneck model the overrepresentation of intermediate- to high-frequency derived alleles is less pronounced with respect to the standard neutral model than compared with (e). As shown in (h), we cannot draw any conclusion on the resulting distribution of D'. When the bottleneck is shifted further into the past, its frequency spectrum resembles the frequency spectrum of a pure expansion model and the resulting distribution of D' becomes applicable again. Shifting the bottleneck further back into the past eventually transforms this model into a standard neutral one.

In conclusion, if a reliable amount of sequenced loci is available such that the observation is anticipated not to deviate too much from the theoretical expectation, one can already foresee the potential of an in-depth analysis for a given dataset from certain tendencies in the frequency spectrum. However, a sample of a relatively small number of loci from an organism that went through a recent bottleneck, as illustrated in Figure 4.4, need not reflect the theoretical expectation, as can be seen in Figure 3.2. Whereas the observed frequency spectrum of the African *D. melanogaster* sample appears entirely smooth, there are stronger fluctuations within the observed European frequency spectrum. In other words, although the recent bottleneck model in Figure 4.4 is similar to some extent to an expansion model in terms of the theoretical frequency spectrum, the observed frequency spectra under both models would appear distinct. We observed that for a given demographic model, i.e. the instantaneous expansion model, that there is a negligible role regarding which parameters we rely on with respect to D', as long as the frequency spectrum provides an appropriate fit to the data. Furthermore, the extension of such a simple demographic model to a model containing multiple changes in population size would not be reasonable regarding a subsequent statistical analysis, if the more complex model does not noticeably improve the fit to the data. If the frequency spectra of a simple and a complex demographic model are virtually identical, test statistics, which are essentially constructed as summary statistics of the site-frequencies, cannot lead to diverse results. By contrast, if a simple model of population size change does not produce an adequate fit to the data, the necessity of incorporating additional demographic parameters becomes apparent; otherwise, a bias is introduced when testing an inappropriately estimated neutral model under variable population size. Interestingly, disregarding the singletons in the African data resulted in a poorer fit of the corresponding frequency spectrum for the revised demographic estimates (LI and STEPHAN 2006) than when taking all the site-frequencies into account. Ideally, demographic estimates would neither be affected nor experience a change in their fit to the observation, when parts of the frequency spectrum are disregarded. This suggests that demographic estimates can be further improved by the

4 Discussion

joint consideration of certain subclasses of derived alleles. Obviously, such an approach demands much more data, distinctly larger sample sizes and computationally exhaustive extensions of previous maximum-likelihood methods. Since the outgroup of $D.$ $simulans$ is used to identify the derived and ancestral variants of a polymorphic site, the possibility remains that misoriented sites introduced a bias (BAUDRY and DEPAULIS 2003). Then again, this effect is mainly observed for sample sizes considerably larger than in both $D.$ $melanogaster$ samples (BAUDRY and DEPAULIS 2003). To avoid this potential problem, the folded frequency spectrum that summarizes mutations of size i and $n-i$ can be applied. However, such an approach appears less promising due to the possible pooling of positively and negatively selected loci into the same class.

When we briefly reconsider the results of Chapter 3 and regard the suggested demographic estimates as sufficient, the number of candidate loci and the corresponding p-values might be seen as too unconvincing to support a major role for strong positive selection in the African sample, with respect to the large number of investigated loci and resultant multiple testing problems. This lack of support may not be surprising due to the conservative approach taken throughout Chapter 3, where we neglected recombination in the simulations for determining the rejection regions of the various test statistics, resulting in a loss of power (cf. WALL 1999). By incorporating the locally estimated recombination rates into the simulations, two loci withstand a conservative Bonferroni-correction with respect to D' and $D'_{-\xi_1}$, respectively. Furthermore, BEISSWANGER and STEPHAN (2008) found evidence for a selective sweep in the region around the polyhomeotic locus, which consists of two tandemly duplicated genes in the African sample.

The in-depth theoretical treatment of neutral non-equilibrium models in this thesis entails future research. The inference of selection as the alternative hypothesis of neutral non-equilibrium models may not be possible with reasonable power, when the underlying demographic estimates are obtained by regarding all loci as putatively neutral. This may lead to a considerable overestimation of the demographic impact and selective influences may be accordingly underestimated. To disentangle the effects of positive and negative selection from fluctuations in population size, we first have to develop a deeper understanding of how these models interact with each other. As already mentioned in Chapter 1, WILLIAMSON et al. (2005) were the first to consider the joint effects of an instantaneous population size change and selection on the frequency spectrum. This work is of particular interest to approach the question of how demography can be separated from weak positive and negative selection. This task appears accomplishable since all loci are equally likely to be subjected to weak selection. In a theoretically more rigorous study, EVANS et al. (2007) have investigated the simultaneous effect of selection and arbitrary changes in population size on the frequency spectrum of derived alleles at independent loci. The development of the non-equilibrium theory for the frequency spectrum of neutral segregating sites, which are partially linked to a beneficial mutation, remains unexplored and represents one of the major prospective goals of modern-day population genetics.

Chapter 4. Discussion

REFERENCES

ACHAZ, G., 2008 Testing for neutrality in samples with sequencing errors. Genetics **179**: 1409–1424.

BARTON, N. H., 1995 Linkage and the limits to natural selection. Genetics **140**: 821–841.

BAUDRY, E. and F. DEPAULIS, 2003 Effect of misoriented sites on neutrality tests with outgroup. Genetics **165**: 1619–1622.

BEISSWANGER, S. and W. STEPHAN, 2008 Evidence that strong positive selection drives neofunctionalization in the tandemly duplicated polyhomeotic genes in Drosophila. Proc. Natl. Acad. Sci. U.S.A. **105**: 5447–5452.

BRAVERMAN, J. M., R. R. HUDSON, N. L. KAPLAN, C. H. LANGLEY, and W. STEPHAN, 1995 The hitchhiking effect on the site frequency spectrum of DNA polymorphisms. Genetics **140**: 783–796.

CHARLESWORTH, B., M. MORGAN, and D. CHARLESWORTH, 1993 The effect of deleterious mutations on neutral molecular variation. Genetics **134**: 1289–1303.

ETHERIDGE, A., P. PFAFFELHUBER, and A. WAKOLBINGER, 2006 An approximate sampling formula under genetic hitchhiking. Ann. Appl. Probab. **16**: 685–729.

EVANS, S. N., Y. SHVETS, and M. SLATKIN, 2007 Non-equilibrium theory of the allele frequency spectrum. Theor Popul Biol **71**: 109–119.

EWENS, W. J., 1972 The sampling theory of selectively neutral alleles. Theor Popul Biol **3**: 87–112.

FAY, J. C. and C.-I. WU, 2000 Hitchhiking under positive Darwinian selection. Genetics **155**: 1405–1413.

FELLER, W., 1951 Diffusion processes in genetics. In *Proc. 2nd Berkeley Symp. on Math. Stat. and Prob.*, pp. 227–246.

REFERENCES

FISHER, R. A., 1930a The distribution of gene ratios for rare mutations. Proc. R. Soc. Edinb. **50**: 205–220.

FISHER, R. A., 1930b *The Genetical Theory of Natural Selection*. Clarendon Press, Oxford.

FU, Y.-X., 1995 Statistical properties of segregating sites. Theoretical Population Biology **48**: 172–197.

FU, Y. X., 1997 Statistical tests of neutrality of mutations against population growth, hitchhiking and background selection. Genetics **147**: 915–925.

FU, Y.-X. and W. H. LI, 1993 Statistical test of neutrality of mutations. Genetics **133**: 693–709.

GILLESPIE, J. H., 1984 The Status of the Neutral Theory: The Neutral Theory of Molecular Evolution. Science **224**: 732–733.

GILLESPIE, J. H., 2000 Genetic drift in an infinite population. The pseudohitchhiking model. Genetics **155**: 909–919.

GILLESPIE, J. H., 2004 *Population genetics: A concise guide* (2nd ed.). Johns Hopkins University Press.

GLINKA, S., L. OMETTO, S. MOUSSET, W. STEPHAN, and D. DE LORENZO, 2003 Demography and natural selection have shaped genetic variation in *Drosophila melanogaster*: a multi-locus approach. Genetics **165**: 1269–1278.

GRIFFITHS, R. C., 1980 Lines of descent in the diffusion approximation of neutral Wright-Fisher models. Theoretical Population Biology **17**: 37–50.

GRIFFITHS, R. C. and S. LESSARD, 2005 Ewens' sampling formula and related formulae: combinatorial proofs, extensions to variable population size and applications to ages of alleles. Theor Popul Biol **68**: 167–177.

GRIFFITHS, R. C. and S. TAVARÉ, 1994 Sampling theory for neutral alleles in a varying environment. Phil. Trans. R. Soc. Lond. B **344**: 403–410.

GRIFFITHS, R. C. and S. TAVARÉ, 1998 The age of a mutation in a general coalescent tree. Stochastic Models **14**: 273–295.

GRIFFITHS, R. C. and S. TAVARÉ, 2003 The genealogy of a neutral mutation. In P. Green, N. Hjort, and S. Richardson (Eds.), *Highly Structured Stochastic Systems, Oxford Statistical Science Series*, Volume 27, pp. 393–412. Oxford University Press.

HADDRILL, P. R., K. R. THORNTON, B. CHARLESWORTH, and P. ANDOLFATTO, 2005 Multilocus patterns of nucleotide variability and the demographic and selection history of *Drosophila melanogaster* populations. Genome Res **15**: 790–799.

HALDANE, J. B. S., 1932 *The Causes of Evolution*. Longmans, Green & Co., London.

HEIN, J., M. H. SCHIERUP, and C. WIUF, 2005 *Gene Genealogies, Variation and Evolution: A Primer in Coalescent Theory*. Oxford University Press.

HERMISSON, J. and P. S. PENNINGS, 2005 Soft sweeps: molecular population genetics of adaptation from standing genetic variation. Genetics **169**: 2335–2352.

REFERENCES

HEY, J., 1999 The neutralist, the fly, and the selectionist. Trends in Ecology and Evolution **14**: 35–38.

HILL, W. G. and A. ROBERTSON, 1968 Linkage disequilibrium in finite populations. Theoretical and Applied Genetics **38**: 226–231.

HUDSON, R. R., 1990 Gene genealogies and the coalescent process. Oxford Surveys in Evolutionary Biology **7**: 1–44.

HUDSON, R. R., 2002 Generating samples under a Wright-Fisher neutral model. Bioinformatics **18**: 337–338.

HUTTER, S., H. LI, S. BEISSWANGER, D. DE LORENZO, and W. STEPHAN, 2007 Distinctly different sex ratios in African and European populations of *Drosophila melanogaster* inferred from chromosome-wide single nucleotide polymorphism data. Genetics **177**: 469–480.

INNAN, H. and W. STEPHAN, 2000 The coalescent in an exponentially growing metapopulation and its application to *Arabidopsis thaliana*. Genetics **155**: 2015–2019.

INTERNATIONAL HAPMAP CONSORTIUM, 2005 A haplotype map of the human genome. Nature **437**: 1299–1320.

JENSEN, J. D., Y. KIM, V. BAUER DUMONT, C. F. AQUADRO, and C. D. BUSTAMANTE, 2005 Distinguishing between selective sweeps and demography using DNA polymorphism data. Genetics **170**: 1401–1410.

JENSEN, J. D., K. R. THORNTON, C. D. BUSTAMANTE, and C. F. AQUADRO, 2007 On the utility of linkage disequilibrium as a statistic for identifying targets of positive selection in nonequilibrium populations. Genetics **176**: 2371–2379.

KAPLAN, N. and R. R. HUDSON, 1985 The use of sample genealogies for studying a selectively neutral m-loci model with recombination. Theor Popul Biol **28**: 382–396.

KAPLAN, N. L., R. R. HUDSON, and C. H. LANGLEY, 1989 The 'hitchhiking effect' revisited. Genetics **123**: 887–899.

KARLIN, S. and J. MCGREGOR, 1972 Addendum to a paper of W. Ewens. Theor Popul Biol **3**: 113–116.

KIM, Y. and R. NIELSEN, 2004 Linkage disequilibrium as a signature of selective sweeps. Genetics **167**: 1513–1524.

KIM, Y. and W. STEPHAN, 2000 Joint effects of genetic hitchhiking and background selection on neutral variation. Genetics **155**: 1415–1427.

KIM, Y. and W. STEPHAN, 2002 Detecting a local signature of genetic hitchhiking along a recombining chromosome. Genetics **160**: 765–777.

KIMURA, M., 1964 Diffusion models in population genetics. Journal of Applied Probability **1**: 177–232.

KIMURA, M., 1968 Evolutionary rate at the molecular level. Nature **217**: 624–626.

KIMURA, M., 1969 The number of heterozygous nucleotide sites maintained in a finite population due to steady flux of mutations. Genetics **61**: 893–903.

REFERENCES

KIMURA, M., 1983 *The Neutral Theory of Molecular Evolution*. Cambridge University Press.

KIMURA, M. and J. F. CROW, 1964 The number of alleles that can be maintained in a finite population. Genetics **49**: 725–738.

KIMURA, M. and T. OHTA, 1978 Stepwise mutation model and distribution of allelic frequencies in a finite population. Proc. Natl. Acad. Sci. U.S.A. **75**: 2868–2872.

KINGMAN, J. F. C., 1982a The coalescent. Stochastic Processes and their Applications **13**: 235–248.

KINGMAN, J. F. C., 1982b On the genealogy of large populations. J. Appl. Prob. **19A**: 27–43.

LI, H. and W. STEPHAN, 2006 Inferring the demographic history and rate of substitution in *Drosophila*. PLoS Genet **2**: e166.

LI, Y. J., Y. SATTA, and N. TAKAHATA, 1999 Paleo-demography of the *Drosophila melanogaster* subgroup: application of the maximum likelihood method. Genes Genet. Syst. **74**: 117–127.

MALÉCOT, G., 1948 *Les mathématiques de l'hérédité*. Masson et Cie, Paris.

MARTH, G. T., E. CZABARKA, J. MURVAI, and S. T. SHERRY, 2004 The allele frequency spectrum in genome-wide human variation data reveals signals of differential demographic history in three large world populations. Genetics **166**: 351–372.

MARUYAMA, T. and P. A. FUERST, 1984 Population bottlenecks and nonequilibrium models in population genetics. I. Allele numbers when populations evolve from zero variability. Genetics **108**: 745–763.

MAYNARD SMITH, J. and J. HAIGH, 1974 The hitch-hiking effect of a favourable gene. Genet. Res. **23**: 23–35.

MCVEAN, G., 2007 The structure of linkage disequilibrium around a selective sweep. Genetics **175**: 1395–1406.

NEI, M., T. MARUYAMA, and R. CHAKRABORTY, 1975 The bottleneck effect and genetic variabiliy in populations. Evolution **29**: 1–10.

NIELSEN, R., S. WILLIAMSON, Y. KIM, M. J. HUBISZ, A. G. CLARK, and C. BUSTAMANTE, 2005 Genomic scans for selective sweeps using SNP data. Genome Research **15**: 1566–1575.

OHTA, T., 1973 Slightly deleterious mutant substitutions in evolution. Nature **246**: 96–98.

OHTA, T., 1976 Role of very slightly deleterious mutations in molecular evolution and polymorphism. Theor Popul Biol **10**: 254–275.

OHTA, T. and M. KIMURA, 1975 The effect of a selected linked locus on heterozygosity of neutral alleles (the hitchhiking effect). Genetical Research **25**: 313–326.

PRZEWORSKI, M., B. CHARLESWORTH, and J. WALL, 1999 Genealogies and weak purifying selection. Mol. Biol. Evol. **16**: 246–252.

REFERENCES

REDON, R., S. ISHIKAWA, K. R. FITCH, L. FEUK, and G. H. PERRY, *et al.*, 2006 Global variation in copy number in the human genome. Nature **444**: 444–454.

ROSENBERG, N. A. and M. NORDBORG, 2002 Genealogical trees, coalescent theory and the analysis of genetic polymorphisms. Nat. Rev. Genet. **3**: 380–390.

SABETI, P. C., D. E. REICH, J. M. HIGGINS, H. Z. P. LEVINE, and D. J. RICHTER, *et al.*, 2002 Detecting recent positive selection in the human genome from haplotype structure. Nature **419**: 832–837.

SAUNDERS, I. W., S. TAVARÉ, and G. A. WATTERSON, 1984 On the genealogy of nested subsamples from a haploid population. Adv. Appl. Prob. **16**: 471–491.

SCHLÖTTERER, C., 2002 A microsatellite-based multilocus screen for the identification of local selective sweeps. Genetics **160**: 753–763.

SCHWEINSBERG, J. and R. DURRETT, 2005 Random partitions approximating the coalescence of lineages during a selective sweep. Ann. Appl. Probab. **15**: 1591–1651.

SLATKIN, M. and R. R. HUDSON, 1991 Pairwise comparisons of mitochondrial DNA sequences in stable and exponentially growing populations. Genetics **129**: 555–562.

SMITH, N. G. and A. EYRE-WALKER, 2002 Adaptive protein evolution in Drosophila. Nature **415**: 1022–1024.

STEPHAN, W., Y. S. SONG, and C. H. LANGLEY, 2006 The hitchhiking effect on linkage disequilibrium between linked neutral loci. Genetics **172**: 2647–2663.

STEPHAN, W., T. H. E. WIEHE, and M. W. LENZ, 1992 The effect of strongly selected substitutions on neutral polymorphism: Analytical results based on diffusion theory. Theoretical Population Biology **41**: 237–254.

TAJIMA, F., 1983 Evolutionary relationship of DNA sequences in finite populations. Genetics **105**: 437–460.

TAJIMA, F., 1989a The effect of change in population size on DNA polymorphism. Genetics **123**: 597–601.

TAJIMA, F., 1989b Statistical method for testing the neutral mutation hypothesis by DNA polymorphism. Genetics **123**: 585–595.

TAKAHATA, N. and M. NEI, 1985 Gene genealogy and variance of interpopulational nucleotide differences. Genetics **110**: 325–344.

TAUTZ, D., 1989 Hypervariability of simple sequences as a general source for polymorphic DNA markers. Nucleic Acids Res. **17**: 6463–6471.

TAVARÉ, S., 1984 Line-of-descent and genealogical processes, and their application in population genetics model. Theoretical Population Biology **26**: 119–164.

WAKELEY, J., 1996 The variance of pairwise nucleotide differences in two populations with migration. Theor Popul Biol **49**: 39–57.

WAKELEY, J., 1997 Using the variance of pairwise differences to estimate the recombination rate. Genet. Res. **69**: 45–48.

REFERENCES

WAKELEY, J., 2008 *Coalescent Theory.* Roberts & Company Publishers, Greenwood Village, Colorado.

WAKELEY, J. and J. HEY, 1997 Estimating ancestral population parameters. Genetics **145:** 847–855.

WALL, J. D., 1999 Recombination and the power of statistical tests of neutrality. Genet. Res. **74:** 65–79.

WATTERSON, G. A., 1975 On the number of segregating sites in genetical systems without recombination. Theoretical Population Biology **7:** 256–276.

WATTERSON, G. A., 1984 Allele frequencies after a bottleneck. Theoretical Population Biology **26:** 387–407.

WIEHE, T., V. NOLTE, D. ZIVKOVIC, and C. SCHLÖTTERER, 2007 Identification of selective sweeps using a dynamically adjusted number of linked microsatellites. Genetics **175:** 207–218.

WIEHE, T. H. and W. STEPHAN, 1993 Analysis of a genetic hitchhiking model, and its application to DNA polymorphism data from *Drosophila melanogaster*. Mol. Biol. Evol. **10:** 842–854.

WILLIAMSON, S. H., R. HERNANDEZ, A. FLEDEL-ALON, L. ZHU, R. NIELSEN, and C. D. BUSTAMANTE, 2005 Simultaneous inference of selection and population growth from patterns of variation in the human genome. Proc. Natl. Acad. Sci. **102:** 7882–7887.

WOLFRAM, S., 1999 *The mathematica book* (4th ed.). Cambridge (UK): Wolfram Media/Cambridge University Press.

WRIGHT, S., 1931 Evolution in Mendelian populations. Genetics **16:** 97–159.

WRIGHT, S., 1938 Size of population and breeding structure in relation to evolution. Science **87:** 430–431.

Die VDM Verlagsservicegesellschaft sucht für wissenschaftliche Verlage abgeschlossene und herausragende

Dissertationen, Habilitationen, Diplomarbeiten, Master Theses, Magisterarbeiten usw.

für die kostenlose Publikation als Fachbuch.

Sie verfügen über eine Arbeit, die hohen inhaltlichen und formalen Ansprüchen genügt, und haben Interesse an einer honorarvergüteten Publikation?

Dann senden Sie bitte erste Informationen über sich und Ihre Arbeit per Email an *info@vdm-vsg.de*.

Sie erhalten kurzfristig unser Feedback!

VDM Verlagsservicegesellschaft mbH
Dudweiler Landstr. 99
D - 66123 Saarbrücken

Telefon +49 681 3720 174
Fax   +49 681 3720 1749

www.vdm-vsg.de

Die VDM Verlagsservicegesellschaft mbH vertritt

Printed by Books on Demand GmbH, Norderstedt / Germany